PROBLEM SOLVED:
BAR MODEL MATH

Bob Krech

New York • Toronto • London • Auckland • Sydney
Mexico City • New Delhi • Hong Kong • Buenos Aires

Editor: Maria L. Chang
Cover design by Tannaz Fassihi
Cover art by Matt Rousell
Interior design by Grafica Inc.
Interior illustrations by Mike Moran

ISBN: 978-0-545-84010-1
Copyright © 2016 by Scholastic Inc.
All rights reserved.
Printed in the U.S.A.
First printing, June 2016.

1 2 3 4 5 6 7 8 9 10   40   22 21 20 19 18 17 16

Table of Contents

Introduction

The very first process standard outlined by both the National Council of Teachers of Mathematics (NCTM) and the more recent Common Core State Standards (CCSS) focuses on understanding problems and "persevering in solving them." They purposefully listed this standard first because problem solving is what teaching math is all about. All of the skills, concepts, knowledge, and strategies students learn in math are basically tools. The eventual goal is that students will apply these math tools to solve problems they encounter in life.

Problem Solved: Bar Model Math is designed to help you and your students learn about a new problem-solving tool—a versatile and effective strategy commonly known as Bar Modeling. A major component of Singapore Math, Bar Modeling has been proven effective in helping students achieve high levels of mathematical competency.

For more than two decades, Singapore's students have consistently ranked among the highest in the world in international math assessments, such as TIMSS (Trends in International Mathematics and Science Study) and PISA (Programme for International Student Assessment). Looking at this success, many schools and districts in the United States and around the world have begun to examine and use ideas found in the Singapore Math curriculum. In my opinion, Bar Modeling is one of the most powerful—if not *the* most powerful—component of Singapore Math.

What Is Bar Modeling?

Bar Modeling is a unique and incredibly versatile strategy that can be used effectively by the youngest elementary school children all the way through to college math majors. This strategy can be applied to a wide range of problem types and contexts. We typically teach children many different strategies (such as Look for a Pattern, Draw a Picture, or Guess and Check) for tackling word problems. With the Bar Modeling method, we can attack any of the various problem types with one singular, powerful approach.

In Bar Modeling, we break out the information in a problem and represent it in a simplified pictorial form using bars or rectangles to represent quantities. The pictorial representation helps children better see and understand the quantities in—and thus the possible solutions to—the problem. The great advantage of the method is that children can visually represent both the given facts in the problem as well as the "unknown" (what they are trying to find out) in a way that allows them to *see* relationships between the quantities, thus promoting flexible thinking about numbers and general number sense. Bar Modeling empowers children to become thinkers, not memorizers.

Where Does Bar Modeling Fit Into the Problem-Solving Process?

When solving word problems, it is always helpful to guide children to using a consistent process. A typical five-step process includes the following:

1. Identify and underline the facts.
2. Identify and circle the question.
3. Eliminate any unnecessary information.
4. Choose a strategy and solve.
5. Go back and check if the answer makes sense.

Bar Modeling falls under the "Choose a strategy and solve" step.

How Does Bar Modeling Work?

When first learning how to solve word problems, children find it very helpful to have problems represented by physical objects. For example, an appropriate problem for primary grades might be: *"Sara has 2 apples. Dana has 4 apples. How many apples do the girls have altogether?"* Instead of using actual apples, children might use physical manipulatives, such as cubes or blocks, to represent the apples. Later, children might draw apples or even cubes to make the representation a little more abstract.

Bar Modeling takes the pictorial representation a step further toward the abstract. Instead of drawing apples or cubes, children represent the quantities by drawing simple rectangles or bars labeled with words, arrows, and numbers. This transition from drawing actual objects to simpler representations is particularly important since quantities eventually become so large that it is impossible to draw or represent them pictorially. A simple rectangle or bar with appropriate labels can do the job.

The first lessons in this book take children step-by-step through this transition. To solve a problem about ladybugs in a garden, for example, children's diagrams would progress as below:

Pictures of ladybugs	🐞 🐞 🐞	3
Ladybug pictures with squares	🐞 🐞 🐞	3
Connected squares	▭▭▭	3
Continuous bar	▭	3

Through each progression, children consistently label each quantity with the amount and, when helpful, its name (e.g., ladybugs, pencils, balls, inches). They follow the same format for labeling the diagrams and using an arrow and question mark to indicate the unknown quantity for which they are solving.

Proving Your Answer/Checking Your Work

But what about children who already have the facts memorized or those who can solve these problems mentally? In these cases, Bar Modeling can serve as a representation or explanation of thinking, a useful way to "prove" or "give evidence" that their answer is correct and makes sense. It is also a second strategy by which children can check their work.

Bar Modeling Variations

The idea of modeling a problem using bars or rectangles can be applied to a wide range of math problems. Within Bar Modeling, there are variations that suit different types of problems. All variations can represent quantities of anything (e.g., apples, miles, dollars, minutes) using labeled bars.

Part-Whole Model

In the Part-Whole Model, two or more parts make up the whole. This model works well for subtraction involving "take away" problems. We can represent the operation of subtraction by crossing out or shading in the quantity being "taken away." The remaining section or "unknown" is indicated by an arrow and a question mark.

Example: *Priya has 8 baseball cards. She gave 5 cards to Stan. How many cards does Priya have left?*

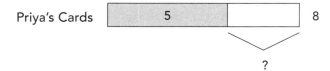

Comparison Model

In the Comparison Model, each quantity in a problem is represented by its own bar. This model has certain advantages when looking closely at relationships between numbers and when breaking up numbers (by place values, for example), so it works particularly well for addition and "comparison" problems.

Addition Example: *Stan has 5 baseball cards. Priya has 8 baseball cards. How many baseball cards do they have altogether?*

The arrow connecting the two numbers shows that the two quantities should be joined or added together to find the unknown, which is indicated by the question mark.

Graphing or Lined Paper

When children work on problems independently (ones not included in this book), it might be helpful to provide them with graph paper or lined paper. Turned horizontally, these types of paper provide lines that children can use as guides for drawing squares and bars.

Comparison Example: *Priya has 8 baseball cards. Stan has 5 baseball cards. How many more baseball cards does Priya have than Stan?*

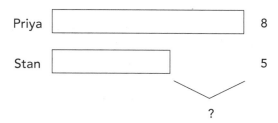

Here, the two bars representing the quantities can be compared easily. The difference between the bars, which is the unknown in this problem, is indicated by the arrow and question mark.

The names of the various models are not so important. Different books and teachers use different names. As you work through the problems in this book, you will find there is much overlap among the methods and there's room for you and your students to create your own variations. Each lesson will detail the actual step-by-step procedures for drawing bar models for these basic variations, particularly in the first two chapters, which are on addition and subtraction.

What Is in This Book?

This book offers a step-by-step approach to helping children learn the Bar Modeling strategy in the context of solving word problems. There are 20 lessons, each with four carefully chosen word problems that gradually increase in difficulty. Use the first two problems in each lesson to introduce the strategy, then have children practice it on their own with the other two problems. All of the word problems are presented in two ways:

- A digital version can be accessed through **www.scholastic.com/ problemsolvedgr2** (registration is required). Display the problems on the interactive whiteboard during your lesson to support your teaching.
- A reproducible version for children can be found in the second part of this book, starting on page 56. These pages feature a lightly printed graph-paper background to help children keep their diagrams neat and organized.

As you work with children on the various word problems, follow the problem-solving process outlined at the top of page 5. Repeating this routine regularly will help children confidently tackle any problem they encounter, especially in testing situations. By the time you have completed the lessons in this book, you will have equipped your students with another powerful tool they can count on as they continue to develop their abilities as excellent, world-class problem solvers.

Silly Problems to Engage Students

The story problems in this book include characters and situations guaranteed to amuse elementary grade children. Silly names (like Fizbop) and ridiculous situations (like a teacher who finds marmots in her students' desks) tend to capture children's interest. Studies have shown that when humor is engaged, attention is engaged. Children will enjoy reading these problems and will look forward to solving them.

Using Pictures and Unit Squares With Addition Problems

As an introduction to the concept of Bar Modeling, children will start by representing the quantities in a word problem with rows of pictures, and then transition to drawing unit squares. They will label the diagrams with a number at the end of each row to indicate the quantity, and a word or abbreviation at the front of each row to identify what the quantity represents. A simple arrow with a question mark will indicate the unknown quantity they are looking for in the problem.

Addition Using Pictures

Materials: student pages 56–57, pencils, projector, interactive whiteboard, markers

Preparation: Distribute copies of pages 56–57 and pencils to children. Go to www.scholastic.com/problemsolvedgr2 and click on Lesson 1. Set up your computer and projector to display the problems on the interactive whiteboard.

Display Problem #1 on the interactive whiteboard.

> Bizzy found 9 bongo beetles. Fizzy found 5 bongo beetles. How many bongo beetles did they find in all?
>
> B. 🐞🐞🐞🐞🐞🐞🐞🐞🐞   9
>
> F. 🐞🐞🐞🐞🐞   5
>
> 〉 ?
>
> 9 + 5 = ___
>
> They found ___ bongo beetles in all.
>
>

Say: *Today we will begin learning about a way to solve word problems called Bar Modeling. It will use some ideas you already know about math and solving problems, as well as some new ideas. Let's take a look at a problem.* Invite the class to read aloud the problem on the board with you.

Ask: *What are the important facts in the problem?* (Bizzy found 9 bongo beetles. Fizzy found 5 bongo beetles.) Have children underline the facts on their paper as you underline them on the board.

Point to the drawings below the word problem on the board.

Say: *Here we see drawings of Bizzy's 9 bongo beetles and Fizzy's 5 bongo beetles. There is a row for each of them, one above the other. How do we know which row is Bizzy's and which row is Fizzy's?* (Their initials are written at the front of the rows.) As children respond to the questions, highlight the answers on the board.

Ask: *How do we know how many bongo beetles are in each row?* (We can count them or look at the number at the end of each row.) *What else do you notice at the end of the row?* (An arrow and a question mark)

Explain: *That arrow joins the two numbers and tells us to find out what those numbers add up to. The question mark is the unknown because we don't know what it is yet. We might also call it the* answer, *or since we're adding, we could also call it the* sum. Highlight the arrow and question mark on the board.

Ask: *So what are we trying to find out?* (How many bongo beetles are there in all?) Circle the question on the board and have children do the same on their papers.

Explain: *So the question, or what we are trying to find out—how many bongo beetles there are in all—is our unknown.*

Point to the answer sentence at the bottom of the board. Invite children to read it aloud with you: "They found *blank* bongo beetles in all."

Explain: *This sentence tells us what we are trying to find out, using words. We will write our answer in the blank in this sentence.* Highlight the answer sentence on the board.

Ask: *What else do you see on the board?* (A number sentence: 9 + 5 = __)

Say: *This number sentence is another way to write what we are doing and trying to find out. But instead of words, we are using numbers. We are joining or adding Bizzy's 9 bongo beetles and Fizzy's 5 bongo beetles to find out how many there are in all. This addition sentence, 9 + 5, shows that we are adding them together.* Highlight the number sentence on the board.

Have children solve the problem together in pairs and fill in the blanks on the number sentence and word sentence on their papers.

Invite children to share their strategies and answers with the class. There are many ways to solve this problem. Some children may count each of the beetles one-by-one. Some may see that they could count by 2s. As children share strategies, diagram their thinking on the whiteboard next to the problem. For example, if a child says, "I noticed that there was 5 + 5 = 10 and then 4 more makes 14," you could illustrate that as shown below.

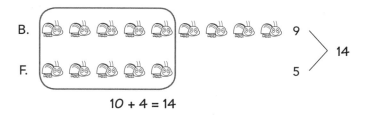

Problem Solved: Bar Model Math (Grade 2) © Scholastic Inc.

Doubles Facts

Many second graders have had experience with basic facts, and of those facts, doubles are among the easiest to remember. So looking for doubles within the bigger addition problem could be a good strategy for many. If children do not come up with it on their own, take some time to share it as a possible strategy for solving one of the problems.

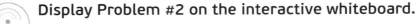

Display Problem #2 on the interactive whiteboard.

Felix saw 4 robo-bunnies hopping down the street.
Then he saw 10 more robo-bunnies join them.
How many robo-bunnies are hopping now?

4 + 10 = ___

There are ___ robo-bunnies hopping now.

Say: *Let's try another problem.* Invite children to read aloud the problem with you. Have children identify the facts and underline them on their papers. Have them circle the question too. Then together, read the answer sentence: "There are *blank* robo-bunnies hopping now."

Ask: *How are the robo-bunnies shown?* (There is a drawing of each robo-bunny. The robo-bunnies are shown in two rows.)

Say: *Putting the robo-bunnies in two rows is a good way to organize the two amounts. It is easier to add and count them than it would be if we put them in one long row.*

Point out how the arrow shows that the two amounts should be joined to find the unknown, which is indicated by the question mark.

Ask: *How can we find out how many robo-bunnies are hopping?* Invite children to suggest strategies, such as counting up 4 from 10 or counting by 2s. Discuss these ideas and diagram them by circling or outlining quantities, as shown below.

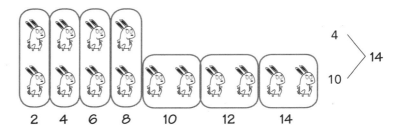

Conclude by writing the answer in the blank and completing the equation. Point out to children that this sum (14) is the same as the sum in Problem #1. Engage children in a discussion about how this is possible since different addends were used.

Have children work on Problems #3 and 4 (page 57) independently or in pairs. Tell children they will need to draw pictures to represent the amounts in each problem. Remind them that they need to line up their rows with each item one above the other. Give children a few minutes to work. Then display each problem on the whiteboard. Call on volunteers to draw and label the diagram on the board. Then invite children to share their strategies for solving the problems, interacting on the whiteboard whenever possible.

Addition Using Pictures and Unit Squares

Materials: student pages 58–59, pencils, projector, interactive whiteboard, markers

Preparation: Distribute copies of pages 58–59 and pencils to children. Go to www.scholastic.com/problemsolvedgr2 and click on Lesson 2. Set up your computer and projector to display the problems on the interactive whiteboard.

In this lesson, children will transition from using pictures to using unit squares as a way to represent quantities. In the whiteboard problems, the pictures will first appear as they did in Lesson 1 and then they will appear with a square drawn around each picture. Each of these squares is a discrete unit called a *unit square*. Student worksheets will display this same discrete unit format. The diagrams will be labeled and an arrow and question mark will indicate the unknown quantity.

Display Problem #5 on the interactive whiteboard.

Boris made 12 brownie bricks on Monday.
He made 4 brownie bricks on Tuesday.
How many brownie bricks did he make altogether?

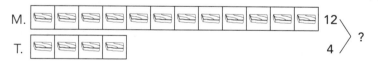

M. [_____] 12
T. [_____] 4 } ?

12 + 4 = ___

Boris made ___ brownie bricks altogether.

Read aloud the problem with the class. Start the problem-solving process by inviting children to identify and underline the facts and circle the question.
Say: *In this problem, we have drawings of the brownie bricks.*

Click on the next page to display a version of the picture that shows a square drawn around each brownie.
Say: *Notice how each brownie now has a square drawn around it. These squares are connected in a row. This makes it easier to count them and to add them.*

Point out how the rows have been labeled *M* and *T*.
Ask: *What do you think* M *and* T *stand for?* (Monday and Tuesday)

Explain: *We can use abbreviations or letters to label the diagram. We don't always have to write out the whole word.*

Invite children to suggest strategies for solving the problem. Common strategies include counting up from the larger number (in this case, 12). In addition to counting up, children might see a 4 + 4 double. Some children might also see an opportunity to find a 10 and then count on the remaining 6.

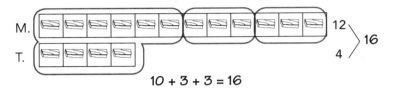

$$10 + 3 + 3 = 16$$

Display Problem #6 on the interactive whiteboard.

There were 3 astronauts at the space station.
The next day, 9 more astronauts joined them.
How many astronauts are at the space station now?

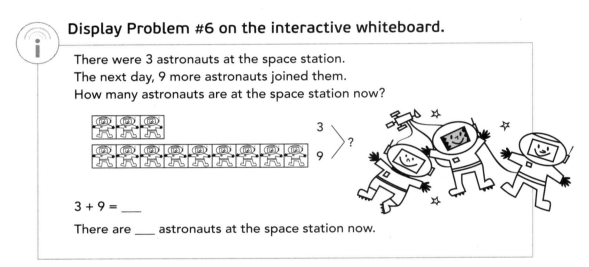

$3 + 9 = $ ___

There are ___ astronauts at the space station now.

Read aloud the problem together. Have children identify the facts and question. Point to the astronauts in the diagram.

Say: *Here again we have drawings of astronauts to help us see the problem. Click on the next page to display another version of the picture on the board.*

Say: *Now each astronaut has a square drawn around it. Notice how the squares are connected in a row, and there are two rows. The squares make it easier for us to count how many astronauts there are and to add them.*

Explain: *Even though the problem mentioned the 3 astronauts first, we don't have to make that the first row of astronauts in our diagram. We could draw them as the second row. Sometimes it is easier to think about a problem if the larger number is shown first. Sometimes it is easier to think about 9 + 3 than 3 + 9.* (See below.)

Invite children to share strategies for solving the problem and to prove their answers are correct. A possible strategy is to count by 2s or even to find a 10 and count up 2 more (see below).

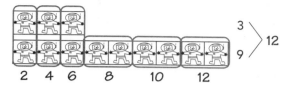

Have children work on Problems #7 and 8 (page 59) independently or in pairs. Remind them to draw either pictures or unit squares to represent the amounts in each problem. Tell children that if they see a problem with more than two amounts (as in #8), they should draw a row of pictures or unit squares for each amount. Give children a few minutes to work. Then display each problem on the whiteboard. Call on volunteers to draw and label the diagram on the board. Then invite children to share their strategies for solving the problems, interacting on the whiteboard whenever possible.

Labeling
Point out to children that they should label quantities in a diagram only when it is helpful. Don't make labeling a requirement for every problem. For example, if a problem compares quantities of carrots and peas, it makes sense to label each quantity to keep them straight. However, if a problem talked about a quantity of pencils, in which some were added or taken away, there's no need to label that quantity since there is nothing to differentiate.

LESSON 3

Addition Using Unit Squares

Materials: student pages 60–61, pencils, projector, interactive whiteboard, markers

Preparation: Distribute copies of pages 60–61 and pencils to children. Go to www.scholastic.com/problemsolvedgr2 and click on Lesson 3. Set up your computer and projector to display the problems on the interactive whiteboard.

As a step toward using bar models, children will draw unit squares around pictures to create discrete units that are easier to count and add. As in previous lessons, diagrams will be labeled with numbers and initials where helpful, and with an arrow and question mark to indicate the unknown quantity.

Display Problem #9 on the interactive whiteboard.

Ned placed 10 peanuts on his head. Ted placed 10 peanuts on his head. How many peanuts did the boys have on their heads altogether?

10 + 10 = ___

The boys had ___ peanuts on their heads altogether.

As always, begin by reading aloud the problem together. Have children start the problem-solving process by identifying and marking the facts and question.
Say: *In our last lesson, we had pictures of the things in each word problem. Then we saw how a square drawn around each picture helped keep the items organized and easy to count. In today's lesson, we will start again with pictures, but this time you will draw the squares around them.*

Model how to draw a square around each peanut on the board, and have children do the same on their papers.
Say: *Notice the gray lines on your paper. They are there to help you draw squares neatly and accurately. The squares will help us count and keep track of each peanut so we don't miss any.*

With the class, count the squares with the peanuts. Note that the number at the end of each row matches the number of squares in that row. Then bring children's attention to the letters *N* and *T* at the beginning of the rows.
Ask: *Why do you think these rows are labeled* N *and* T? (To show that one row is for Ned and one row is for Ted)

Next, point to the arrow and question mark at the end of the rows.
Say: *The arrow tells us that we're joining the two rows to find the answer or the unknown, which is the question mark.*

Have children work in pairs to share strategies for finding the unknown. Many children will know the 10 + 10 doubles fact, while others may use the two rows to count by 2s to get to 20.

Display Problem #10 on the interactive whiteboard.

Billy Worm counted 13 birds. Willy Worm counted 5 birds. How many birds did they count altogether?

13 + 5 = ___ birds

They counted ___ birds altogether.

Read aloud the problem together and have children identify and mark the facts and the question on their papers.
Say: *Again we have pictures that show how many birds are there. But this time there are no numbers, labels, or unknowns. On your paper, add these labels to your diagrams. Start by drawing a square around each bird to make the birds easier to count.* Model this on the board.
Say: *Let's count how many squares are in each row, then write the number at the end of each row. Make sure to line up those numbers, one above the other, even though the rows are not the same length.*
Continue: *Now let's label the rows. How should we do that?* (Write *B* for Billy and *W* for Willy.) Model this on the board as children do the same on their papers.

Say: *Now we have to show where our unknown is. How do we do that?* (Draw an arrow joining the two numbers and write a question mark.) Do this on the board and check that children have done the same.

Invite children to share strategies for solving the problem. Helpful strategies include finding the 5 + 5 double and outlining it, as well as counting by 2s.

Have children work on Problems #11 and 12 (page 61) independently or in pairs. Remind them to draw either pictures or unit squares to represent each problem. Give children a few minutes to work. Then display each problem on the whiteboard. Call on volunteers to draw and label the diagram on the board. Then invite children to share their strategies for solving the problems, interacting on the whiteboard whenever possible.

> **"Represent"—
> A Key Word**
>
> Throughout this book we tell children to "represent" quantities with drawings and symbols. This is a major idea in math—we don't always need the physical objects to count, add, subtract, or do any other operation. Instead we can use drawings or symbols to *represent*, or stand for, amounts. Tell children: *In math we often use a simple drawing or symbol to represent or stand for something that might be too hard or take too long to draw. A square can represent a cat, a dog, a cookie, a rocket — anything you want.*

LESSON 4

Drawing Unit Squares to Add

Materials: student pages 62–63, pencils, projector, interactive whiteboard, markers

Preparation: Distribute copies of pages 62–63 and pencils to children. Go to www.scholastic.com/problemsolvedgr2 and click on Lesson 4. Set up your computer and projector to display the problems on the interactive whiteboard.

In this lesson, children will see that squares without pictures inside can still represent the quantities in a problem effectively. Children will read the problem and draw their own squares to represent the quantities. They will also label their diagrams with numbers, when helpful, and with an arrow and question mark to indicate the unknown quantity.

Display Problem #13 on the interactive whiteboard.

Mandy sold 5 boxes of candy to the groundhogs.
She sold 11 boxes of candy to the squirrels.
How many boxes of candy did Mandy sell in all?

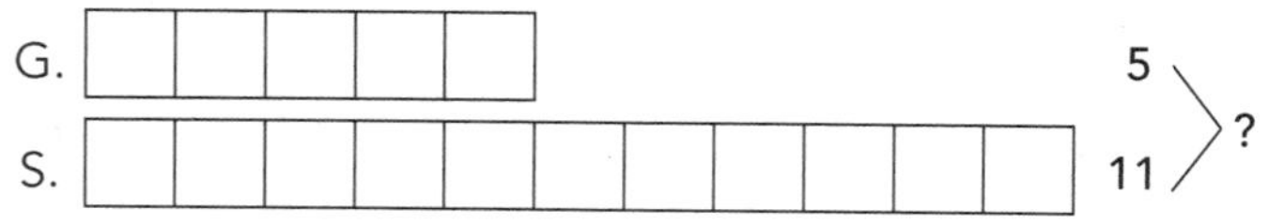

G. ▢▢▢▢▢   5 ⟩
S. ▢▢▢▢▢▢▢▢▢▢▢   11 ⟩ ?

5 + 11 = ___

Mandy sold ___ boxes of candy in all.

Begin as always by reading the problem aloud together. Have children start the problem-solving process by identifying the facts and the question. Point out that the diagram shows only squares with no pictures inside them. Invite children to count the squares with you.

Say: *A square is so much easier and quicker to draw than a box of candy.*

Drawing just squares also makes it easier to count how many boxes of candy there are and to add them all up.

Direct children to look at the problem on their papers. Point out that the squares are not labeled. Have them label the diagram with numbers, letters, and an arrow and question mark to indicate the unknown.

Ask: *How should you label the rows?* (Label the row with 5 squares *G* for groundhogs and the other row *S* for squirrels.) Remind children that when labeling the amounts, they can just use an initial, like *G* and *S*. Check to make sure children label their diagrams correctly, then call on a volunteer to label the diagram on the board.

Invite children to work in pairs to share strategies for solving the problem. Strategies might include looking for doubles, such as 5 + 5, 6 + 6, 7 + 7, or 8 + 8. The higher the double used, the easier the problem, as there is very little to add on to the doubles sum. Counting by 2s also makes sense here, as does finding a 10 and then adding on 6 more. Show these ideas by illustrating on the board.

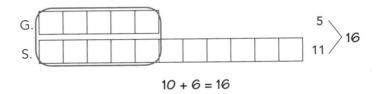

10 + 6 = 16

Display Problem #14 on the interactive whiteboard.

6 bats flew in the kitchen at night. In the morning, 7 more bats joined them. Then during the afternoon, 3 more bats flew in. How many bats are in the kitchen altogether?

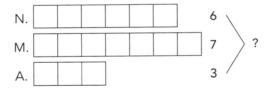

6 + 7 + 3 = ___

There are ___ bats in the kitchen altogether.

Read aloud the problem together. Then have children identify and mark the facts and question on their papers.

Say: *Notice that we have three quantities in this problem. How are we going to show these quantities?* (Draw three rows of unit squares.) *On your papers, draw the squares to represent all the bats. Remember, the squares do not have to be perfect. You do not need a ruler. Just draw a simple square for each bat. The lines on your sheet should help. Also, remember to make a row for each amount in the problem. Make sure the squares are lined up one above the other.* Give children a few minutes to complete their drawings.

Ask: *How many rows of squares did you draw?* (3) *How many squares did you draw for night?* (6) *How many for morning?* (7) *How many for the afternoon?* (3)

Call on volunteers to draw the diagram on the board, keeping the squares connected in rows, one row above the other. Have children label the rows with numbers and with letters for night, morning, and afternoon. Remind them to show the arrow that joins the quantities and the question mark to show the unknown.

Then invite children to share strategies for solving the problem. Some children may see that 7 + 3 can be combined to make a 10, and adding 6 to 10 makes 16. Another strategy is to see the 6 + 6 double and then add on 4 more to get 16.

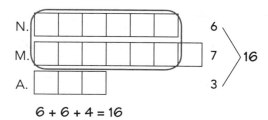

6 + 6 + 4 = 16

Have children work on Problems #15 and 16 (page 63) independently or in pairs. Remind them to draw unit squares to represent the amounts in each problem. Give children a few minutes to work. Then display each problem on the whiteboard. Call on volunteers to draw and label the diagram on the board. Then invite children to share their strategies for solving the problems, interacting on the whiteboard whenever possible.

Homework

If you want children to practice these strategies at home, assign developmentally appropriate word problems from any source. Ask children to solve the problems using the Bar Modeling method that you have been practicing in class. Provide graph paper or horizontally lined paper to help children with their drawings.

Using Pictures and Unit Squares With Subtraction Problems

Subtraction is about finding the difference between two quantities. Sometimes we begin with a quantity and take away from that. For example: *Bob had 4 cookies. He ate 2 cookies. How many cookies are left?* Other times we compare two quantities: *Karen had 6 cookies. Bob had 4 cookies. How many more cookies did Karen have than Bob?*

Subtracting and adding up are both reasonable strategies for finding the answer, but either way we are finding the difference between the two quantities. Bar Modeling is a great way to help children visualize, represent, and understand subtraction—in both "take away" and "comparison" situations.

Follow the routine of reading aloud problems together, going through the problem-solving process (identifying and underlining the facts and circling the question), and sharing and discussing solutions. As much as possible, have children work in pairs.

LESSON 5

"Take Away" Problems
Using Pictures and Unit Squares

Materials: student pages 64–65, pencils, projector, interactive whiteboard, markers

Preparation: Distribute copies of pages 64–65 and pencils to children. Go to www.scholastic.com/problemsolvedgr2 and click on Lesson 5. Set up your computer and projector to display the problems on the interactive whiteboard.

The focus of this lesson is basic "take away" subtraction problems. As with addition, the process starts with simple drawings representing the quantities in the problem. Then the drawings will appear with a square drawn around each picture, making each unit more discrete.

Display Problem #17 on the interactive whiteboard.

Pez had 9 noodles. The Goo-Goo Monster ate 5 of them.
How many noodles does Pez have left?

9 − 5 = ___

Pez has ___ noodles left.

Ask: *What are the important facts in this problem?* (Pez had 9 noodles. The Goo-Goo Monster ate 5 of them.)

Say: *On your paper, you will notice all of the noodles are drawn in one row. The number at the end of the row tells us how many there are in that row. Some of the noodles are crossed out.*

Ask: *What do you notice about the arrow and question mark pointing to the unknown? How are they different from when we've worked on addition problems?* (The arrow and question mark point to the remaining noodles that have not been eaten. Instead of joining things together, they show what is left when some things have been removed.)

Next, ask children to identify the question and circle it on their papers. Then have them find the answer sentence and read it together: "Pez has *blank* noodles left."

Ask: *So what are we trying to find out?* (How many noodles are left?) *How can we show this on our diagram?* (Cross out the noodles the Goo-Goo Monster ate.)

On the whiteboard, model crossing out the 5 noodles that were eaten. (Simply draw over the work already done.)

Explain: *We see 4 noodles that are not crossed out. These are the noodles that are left. The arrow with the question mark shows that this is the unknown quantity.*

Invite children to share strategies and check their answers. Some children will immediately recognize there are 4 left, while others may have to count the remainder by 1s or 2s. Fill in the answer in the sentence and in the equation.

Crossing Out—Does Direction Matter?

When subtracting, explain to children that it doesn't matter where they begin crossing out or shading in the row of pictures or squares. Starting at the front or back of the row will yield the same results. You may want to demonstrate this a couple of times to show that the results are the same. Then let children use their personal preference.

Display Problem #18 on the interactive whiteboard.

There were 12 flying turtles. Then 9 flew away.
How many flying turtles were left?

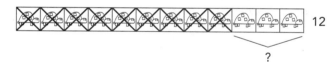 12

?

12 − 9 = ___

There were ___ flying turtles left.

Invite children to read the problem aloud. Then have them identify and underline the facts and circle the question on their papers.

Say: *We can see all the turtles drawn out in one row. But there's an easier way to show the turtles without having to draw each one.*

Click on the next page to display a version of the picture that shows a square around each turtle.

Say: *Notice how each turtle now has a square drawn around it, making it easier to count them all.*

Ask: *So how many flying turtles are there at the start of the problem?* (12) *How many turtles flew away?* (9) *How can we show that they are gone?* (Cross out or shade in the squares.)

Model crossing out 9 turtles on the board and have children do the same on their papers.

Ask: *How do we know what is left?* (We can count the turtles that are not crossed out.)

Guide children to notice the "unknown" as indicated by the arrow and question mark pointed at the remaining turtles.

Say: *Here is our unknown—the number of flying turtles left.*

Since there are only 3 flying turtles left, it is fairly easy for children to count the remainder. Conclude by filling in the number in the answer blank and in the equation.

Have children work on Problems #19 and 20 (page 65) independently or in pairs. Remind them to draw either pictures or unit squares to represent the amounts in each problem and to cross out items as they subtract. Make sure they show the unknown quantity with an arrow and question mark. Give children a few minutes to work. Then display each problem on the whiteboard. Call on volunteers to draw and label the diagram on the board. Then invite children to share their strategies for solving the problems, interacting on the whiteboard whenever possible.

"Comparison" Problems Using Pictures and Unit Squares

Materials: student pages 66–67, pencils, projector, interactive whiteboard, markers

Preparation: Distribute copies of pages 66–67 and pencils to children. Go to www.scholastic.com/problemsolvedgr2 and click on Lesson 6. Set up your computer and projector to display the problems on the interactive whiteboard.

This lesson focuses on "comparison" subtraction problems. Children will continue to work with pictures both with and without unit squares around them.

Display Problem #21 on the interactive whiteboard.

Clancy has 14 blue shoes. He also has 8 purple shoes. How many more blue shoes are there than purple shoes?

B. 14

P. 8

?

14 – 8 = ___

There are ___ more blue shoes than purple shoes.

Read aloud the problem together. Then ask children to identify and underline the facts and circle the question on their papers.

Explain: *Here is our first comparison problem. Even though nothing is being taken away, we can still use subtraction to solve this problem. Since there are two different things in this problem—blue shoes and purple shoes—we will represent them both, one group above the other, like we do when adding.*

Ask: *So which are there more of—blue shoes or purple shoes?* (Blue shoes)

Say: *Right. We can see there are more blue shoes in the picture. We want to find out how many more.*

Guide children to notice the arrow pointing to the area at the end of the second row. The question mark indicates two ideas simultaneously: It shows that there are 6 more blue shoes and 6 fewer purple shoes. Emphasize this by drawing a line at the end of the 8 purple shoes through the row of blue shoes, as shown below. Demonstrate this and tell children how this can help us see the difference between the two quantities.

Ask children to find an answer, share their strategies with partners, and then share back solutions with the class.

Some solutions might include connecting blue shoes and purple shoes in one-to-one correspondence with lines or counting the leftover blue shoes that don't match up with purple shoes. In either case, help children understand that this is finding the difference and we can express it with a subtraction sentence.

Display Problem #22 on the interactive whiteboard.

Chef Crumb baked 2 spinach cakes. She baked 17 cucumber cakes. What is the difference between the number of cucumber cakes and spinach cakes?

17 – 2 = ___

The difference between the number of cucumber cakes and spinach cakes is ___.

This comparison problem uses different wording in the question, specifically asking us to identify the difference between the quantities. The wording may be unfamiliar to children, so guide them in a discussion about what is being asked.

Ask: *What is the problem asking us to find?* (The difference between the number of cucumber cakes and spinach cakes) *What's another way to ask the same question?* (How many more cucumber cakes are there than spinach cakes?)

Point to the two rows of cakes drawn on the interactive whiteboard. Then click on the page to display another version of the picture that shows a square around each cake.

Say: *Again, we see a square around each cake, making it easier to count them. Notice how the squares are lined up one-to-one in two rows, one above the other for easy comparison.*

Invite children to share and demonstrate strategies for solving the problem. As in most comparison problems, drawing a line between the two quantities helps emphasize the difference between them (see below).

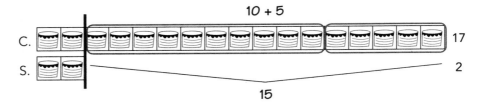

Drawing the line can help children easily see and count the difference by 1s or 2s. Some children also might think about how counting back 2 from 17 will result in 15.

Have children work on Problems #23 and 24 (page 67) independently or in pairs. Remind them to draw either pictures or unit squares to represent the amounts in each problem, making sure to show the unknown quantity with an arrow and question mark. Give children a few minutes to work. Then display each problem on the whiteboard. Call on volunteers to draw and label the diagram on the board. Then invite children to share their strategies for solving the problems, interacting on the whiteboard whenever possible.

Subtraction Using Unit Squares

Materials: student pages 68–69, pencils, projector, interactive whiteboard, markers

Preparation: Distribute copies of pages 68–69 and pencils to children. Go to www.scholastic.com/problemsolvedgr2 and click on Lesson 7. Set up your computer and projector to display the problems on the interactive whiteboard.

This lesson continues with basic subtraction, modeling both "take away" and "comparison" problems. But instead of using pictures, children will now use only squares to represent the quantities.

Display Problem #25 on the interactive whiteboard.

Henrietta Hen will sing 20 songs at her concert. She has sung 10 songs so far. How many more songs does she have to sing?

20

?

20 – 10 = ___

Henrietta has ____ more songs to sing.

Read aloud the problem together. Then have children identify and underline the facts and circle the question on their papers.

Explain: *Now that we know how to use this strategy for subtracting, we will remove the drawings and use just the squares. Drawing so many things can be hard, but squares are easier to draw and can still represent the amounts of different things in the problem.*

Ask: *In this problem, what does each square represent or stand for?* (A song)

Say: *You can see all these squares are in a row. We are not comparing anything. We have some songs that were sung and some that were not. How can we show this?* (Cross out or shade in the songs that were sung.)

Model crossing out or shading in the squares on the whiteboard, then ask children to share strategies for finding the answer. Some may decide to count the remainder by 2s, while others may count backwards 10 from 20. Many will know the 10 + 10 doubles fact. The representation serves as proof or evidence that the fact is true.

Display Problem #26 on the interactive whiteboard.

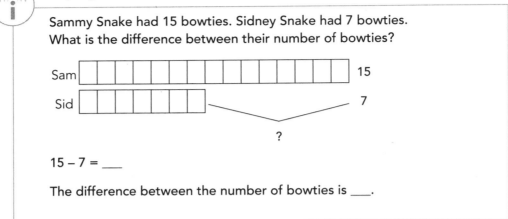

Sammy Snake had 15 bowties. Sidney Snake had 7 bowties.
What is the difference between their number of bowties?

15 – 7 = ___

The difference between the number of bowties is ___.

Read aloud the problem and ask students to identify the facts and question. **Ask:** *Why are the squares in two groups, one above the other?* (Because there are two different groups of bowties, one for Sammy's and one for Sidney's. We are comparing them.)

Point out the labels and where the arrow and question mark are placed to indicate the unknown quantity. Emphasize how these labels and markings help us see clearly the facts of the problem.

Guide children in comparing and counting the difference between the two quantities. Draw a line at the end of Sidney's row through Sammy's row (as in the previous lesson) to highlight the comparison. The difference may be determined efficiently by counting by 2s or finding doubles, such as 3 + 3 or 4 + 4.

Have children work on Problems #27 and 28 (page 69) independently or in pairs. Remind them to draw unit squares to represent the amounts in each problem, making sure to show the unknown quantity with an arrow and question mark. Give children a few minutes to work. Then display each problem on the whiteboard. Call on volunteers to draw and label the diagram on the board. Then invite children to share their strategies for solving the problems, interacting on the whiteboard whenever possible.

Drawing Unit Squares to Subtract

Materials: student pages 70–71, pencils, projector, interactive whiteboard, markers

Preparation: Distribute copies of pages 70–71 and pencils to children. Go to www.scholastic.com/problemsolvedgr2 and click on Lesson 8. Set up your computer and projector to display the problems on the interactive whiteboard.

Children will draw their own unit squares to represent the quantities in the problems in this lesson. They will also label the squares with numbers and letters when necessary, and with an arrow and a question mark for the unknown quantity.

Display Problem #29 on the interactive whiteboard.

Fantastic Freda had 16 pumpkin pops.
She gave 7 to her sister Fluffy.
How many pumpkin pops does Freda have left?

16

?

16 – 7 = ___

Freda has ___ pumpkin pops left.

Read aloud the problem together with the class. Then ask children to identify and mark the facts and question on their papers.

Say: *You will need to draw squares to represent the pumpkin pops in the problem.*

Ask: *How many pumpkin pops does Freda have at the start of the problem?* (16) *How can we show or represent this?* (Draw a square for each pumpkin pop.)

Call on a volunteer to draw 16 squares linked together on the board and label the quantity at the end of the row. Have children do the same on their papers.

Say: *So Freda started out with 16 pops. Then what happened?* (She gave 7 to her sister.) *How can we show this?* (Cross out or shade in 7 squares.)

Call on another volunteer to cross out the quantity on the board. Have the rest of the class do the same on their papers.

Ask: *What are we trying to find out?* (How many pumpkin pops are left?)

Guide children to draw an arrow and a question mark to show where this unknown quantity is on the diagram. Model this on the board while others follow on their papers. Then invite children to share strategies for counting the remaining pops.

Display Problem #30 on the interactive whiteboard.

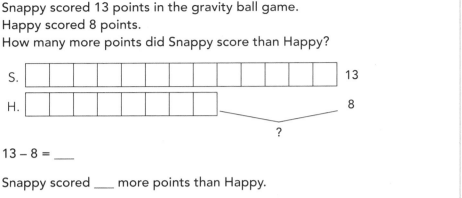

Snappy scored 13 points in the gravity ball game.
Happy scored 8 points.
How many more points did Snappy score than Happy?

S. ⬚⬚⬚⬚⬚⬚⬚⬚⬚⬚⬚⬚⬚ 13

H. ⬚⬚⬚⬚⬚⬚⬚⬚ 8
?

13 – 8 = ___

Snappy scored ___ more points than Happy.

Read aloud the problem together.

Ask: *How should we represent the points in this problem?* (Draw a row of squares for Snappy's points and another row of squares underneath for Happy's points.) *Why do we need to show two rows of squares?* (Because we have two separate quantities that we are comparing.)

Have children draw the rows of squares on their papers, instructing them to label their diagrams with initials to show which row of squares belong to whom, quantities at the end of each row, and an arrow and question mark to indicate the unknown. Then call on volunteers to share their diagrams on the board.

Invite children to share their strategies for solving the problem. One effective strategy, for example, is to draw a line from the end of the 8 row up to the 13 row to show where the 8 in the first row matches up.

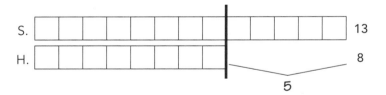

S. ⬚⬚⬚⬚⬚⬚⬚⬚│⬚⬚⬚⬚⬚ 13

H. ⬚⬚⬚⬚⬚⬚⬚⬚ 8
5

Have children work on Problems #31 and 32 (page 71) independently or in pairs. Remind them to draw unit squares to represent the amounts in each problem, making sure to show the unknown quantity with an arrow and question mark. Give children a few minutes to work. Then display each problem on the whiteboard. Call on volunteers to draw and label the diagram on the board. Then invite children to share their strategies for solving the problems, interacting on the whiteboard whenever possible.

Bar Modeling With Addition and Subtraction (Within 100)

As we move to addition and subtraction problems with larger numbers, children will transition from drawing unit squares to continuous bar models. Emphasize the importance of properly drawing and labeling diagrams, including the arrow and question mark that indicates the unknown quantity.

Continue to follow the same lesson routine established earlier: read aloud problems together, go through the problem-solving process (identifying and underlining the facts and circling the question), and share and discuss solutions. Have children work in pairs as much as possible.

LESSON 9

Addition Using Bar Models

Materials: student pages 72–73, pencils, projector, interactive whiteboard, markers

Preparation: Distribute copies of pages 72–73 and pencils to children. Go to www.scholastic.com/problemsolvedgr2 and click on Lesson 9. Set up your computer and projector to display the problems on the interactive whiteboard.

This lesson begins with a problem that shows unit squares linked together. The unit squares are then removed so children see that continuous bars can be used to represent these quantities. This is an important transition. Point out that as numbers get larger, it becomes more difficult and time-consuming to draw so many squares; a continuous bar works more easily. Remind children that bars should be labeled with numbers and initials (when helpful), as well as with an arrow and question mark for the unknown quantity.

Display Problem #33 on the interactive whiteboard.

Mr. Cat R. Pillar has 22 socks. He found 20 more under the bed. How many socks does Mr. Pillar have now?

22
20
?

22 + 20 = ___

Mr. Pillar has ___ socks now.

Smaller Problems Inside Bigger Problems

When children encounter a problem where they are adding quantities, such as 40 + 30, you may want to remind them to look for smaller problems within the bigger problem. For example, if we know that 4 + 3 = 7, we also know that 40 + 30 = 70. This strategy can be used in any part of an equation—ones, tens, hundreds, and so on.

Read aloud the problem together and ask children to identify the facts and question. Note that we have moved from drawing socks to drawing unit squares to represent the socks in the problem. Ask children to draw unit squares to represent the quantities on their papers.

Say: *That's a lot of unit squares you have to draw to show all those socks. What if we were to do this instead?* Click on the next page to display a version of the diagram that shows continuous bars with the interior lines erased (see below).

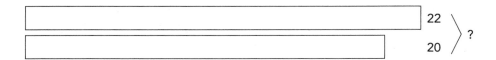

Explain: *Notice how they look like bars or rectangles. As we work with larger numbers, it will be easier if we draw a bar for each amount instead of drawing so many little squares. We can see that one bar is a little longer than the other. That makes sense because 22 is a little more than 20. Always remember to label the bars with their values.*

Ask: *Now, since we can't count squares anymore, how can we use these bars to solve the problem?*

This is a key moment. Diagram students' strategies, and if none are offered, demonstrate the following strategy on the whiteboard.

Explain: *We can always draw a line to cut up or section a bar to help us see and remember how much a certain part of a bar is worth. For example, we can draw a line that shows where 20 on the top bar matches up with 20 on the bottom bar. We know 20 + 20 = 40, so 2 more gives us 42.*

When using an approach like this, it is helpful to show children they can label the two sections of the top bar with numbers, as shown below.

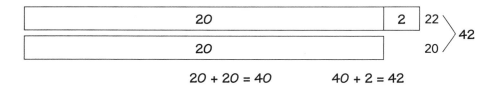

Display Problem #34 on the interactive whiteboard.

Spot chased 40 elephants. He chased 24 giraffes.
He chased 32 rhinos. How many animals did Spot chase in all?

E. [_____] 40

G. [_____] 24 } ?

R. [_____] 32

40 + 24 + 32 = ___

Spot chased ___ animals in all.

Point out that this problem has three addends. Explain that this simply means we need to draw three bars instead of two.

Say: *Drawing unit squares to represent all these animals would take too long. We are using numbers now, like 40, that are larger than the number of squares on your paper. So let's draw bars instead.*

Call on volunteers to draw the bars on the board. Guide them to draw the lengths of the bars in relation to one another and to label the bars with their values and abbreviations for the animals' names. Then draw an arrow and question mark to show the unknown. Have the rest of the class do the same on their papers.

Explain: *Notice that the elephants' bar is the longest and the giraffes' bar is the shortest. You don't always have to draw the bars in the order they appear in the problem. You can move them around if it makes your adding easier. For example, some people prefer to draw them in order from longest to shortest* (see diagram on next page). *Can you see why?* Encourage children to suggest ideas.

Say: *So we have to add three amounts. What are some ways to do this?* Invite children to share strategies on the board to demonstrate their thinking.

Explain: *Remember we can always draw lines to cut up or section the bars. For example, we can look at the elephants' bar and section it in half, so we have 20 and 20. Then we can draw a line down to the rhinos' bar to match the 20 from the elephants'. What is the rest of the rhinos' bar worth?* (12) *We can also section the rest of the rhinos' bar to 10 and 2; that will help us when we add all the tens later. Then we can also section the giraffes' bar to 20 and 4. Drawing these lines helps us see that we can add the first section of each bar: 20 + 20 + 20 = 60. Then we can add the remaining tens: 20 + 10 = 30. Now we have 90. Lastly, we can add up the ones: 2 + 4 = 6. And so we have the final total: 90 + 6 = 96.*

E.	20	20		40
R.	20	10	2	32
G.	20	4		24

96

$$20 + 20 + 20 = 60 \qquad 2 + 4 = 6$$
$$20 + 10 = 30 \qquad 90 + 6 = 96$$
$$60 + 30 = 90$$

Bars Don't Have to be Exact

When children begin to draw bars, they will probably use the graph paper squares to make exactly numbered bars. Help children understand a bar that is 2 inches or 8 squares long could represent the quantity 100. But they should try to draw bars in the same problem in relation to each other.

For example, if a problem has the amounts 78 and 32, ask children: *Is 32 about half of 78? A quarter of 78?* This promotes good number sense and mental math practice. Help children understand that the 32 bar should be a little less than half of the 78 bar.

Explain: *So with larger numbers it is usually easier to draw bars instead of unit squares. We can put the bars in any order we want. We can also draw lines to section the bars to help us add easier numbers. Just remember to write the value for each section as you go along.*

Have children work on Problems #35 and 36 (page 73) independently or in pairs. Remind them to draw continuous bars to represent the amounts in each problem. Give children a few minutes to work. Then display each problem on the whiteboard. Call on volunteers to draw and label the diagram on the board. Then invite children to share their strategies for solving the problems, interacting on the whiteboard whenever possible.

LESSON 10

Addition With Regrouping

Materials: student pages 74–75, pencils, projector, interactive whiteboard, markers

Preparation: Distribute copies of pages 74–75 and pencils to children. Go to www.scholastic.com/problemsolvedgr2 and click on Lesson 10. Set up your computer and projector to display the problems on the interactive whiteboard.

The problems in this lesson involve addition that normally requires regrouping in the traditional algorithm.

Display Problem #37 on the interactive whiteboard.

Ludmilla jumped 39 times over her brother Boris.
She jumped 61 times over her sister Olga.
How many times did Ludmilla jump altogether?

39
61
?

39 + 61 = ___

Ludmilla jumped ___ times altogether.

Read aloud the problem together. Then ask children to identify and mark the facts and the question on their papers.

Say: *Once again we're dealing with large numbers, so let's draw bars to show all those jumps. Draw two bars on your paper, one above the other—one for 39 jumps and another for 61 jumps. About how long will the second bar be?* (About a third longer than the first bar)

Remind children to label the rows with numbers (and initials, if helpful) and indicate the unknown with an arrow and question mark. Call on a volunteer to model this on the board.

Explain: *Without unit squares to count we need another strategy to help us solve this problem. One strategy would be to section the top bar into 30 + 9 and the bottom bar into 60 + 1. We can then combine 30 + 60 to make 90, then add on 9 more to make 99, and then add the final 1 to make 100.*

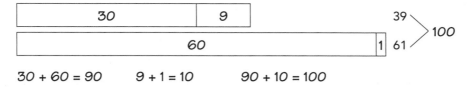

$$30 + 60 = 90 \qquad 9 + 1 = 10 \qquad 90 + 10 = 100$$

This is a good problem to use when talking to children about finding 10s to make addition easier. After we combine 30 and 60 to make 90, we have 9 + 1 left, which equals 10. 90 + 10 = 100.

You will find that these pictorial strategies help support and build children's mental math skills.

Display Problem #38 on the interactive whiteboard.

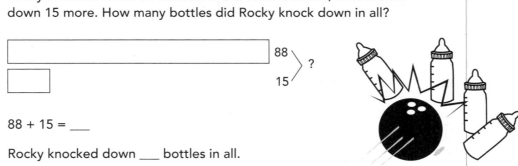

In the first round of the Baby Bottle Bowling Championship, Rocky knocked down 88 bottles. In the second round, he knocked down 15 more. How many bottles did Rocky knock down in all?

88 + 15 = ___

Rocky knocked down ___ bottles in all.

Read aloud the problem together and go through the problem-solving process with the class.

Say: *We're dealing with large numbers again, so we need to draw bars to show all those bottles. Draw two bars on your paper, one above the other—one for the 88 bottles and another for the 15 bottles. About how long should the second bar be?* (Much smaller than the first bar. About 1/5 is reasonable, but remind children again that these are just approximations.)

Reference Chart

To support student understanding, create a reference chart for Bar Modeling to display somewhere in the classroom. On the chart consider including diagrams, vocabulary, and examples of different types of problems and strategies.

Make sure children label the rows with their values and indicate the unknown with an arrow and question mark. Call on a volunteer to model this on the board.

Explain: *Let's see if we can find some easy numbers to add in this problem. We can start by sectioning the top bar into 80 + 8 and the bottom bar into 10 + 5.*

Some children will know to add 80 + 10 = 90 and then add on the 8 to make 98. They could then count up 5 more to make 103.

To make it even easier, have children think about the 5 section of the 15 bar. Challenge them to find another 5 in the top bar to make another 10. (They can section the 8 in the top bar into a 5 and 3.) Children could then add 5 + 5 = 10, which added to 90 makes 100. Then add on the remaining 3 to make 103.

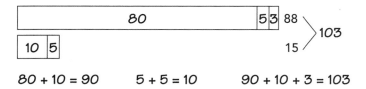

$$80 + 10 = 90 \qquad 5 + 5 = 10 \qquad 90 + 10 + 3 = 103$$

Invite children to share other possible strategies on the board and discuss.

Have children work on Problems #39 and 40 (page 75) independently or in pairs. Remind them to draw bars to represent the amounts in each problem. Give children a few minutes to work. Then display each problem on the whiteboard. Call on volunteers to draw and label the diagram on the board. Then invite children to share their strategies for solving the problems, interacting on the whiteboard whenever possible.

LESSON 11

Subtraction Using Bar Models

Materials: student pages 76–77, pencils, projector, interactive whiteboard, markers

Preparation: Distribute copies of pages 76–77 and pencils to children. Go to www.scholastic.com/problemsolvedgr2 and click on Lesson 11. Set up your computer and projector to display the problems on the interactive whiteboard.

The subtraction problems in this lesson involve larger numbers. Children will review how to subtract with unit squares then quickly transition to continuous bars. Remind children that as numbers get larger, it becomes more difficult and time-consuming to draw so many squares.

Display Problem #41 on the interactive whiteboard.

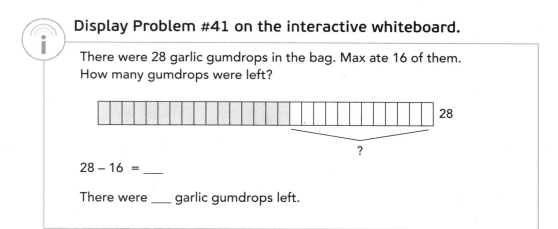

There were 28 garlic gumdrops in the bag. Max ate 16 of them.
How many gumdrops were left?

28 – 16 = ___

There were ___ garlic gumdrops left.

Read aloud the problem together. Ask children to identify and mark the facts and question on their papers.

Say: *In today's lesson we will see subtraction problems with larger numbers. Let's start with this problem on the board. We see a row of connected squares. First, let's label this diagram and draw an arrow and question mark to show the unknown.* Call on a volunteer to model this on the board.

Ask: *How can we solve this problem using our diagram?* (Cross out or shade in 16 of the 28 squares, then count the remaining squares.)

Say: *Notice how there's a lot of unit squares to draw. As the numbers in our problems get bigger, it will be easier if we can simply draw a continuous bar, like we did when adding larger numbers.* Click on the next page to display this alternative diagram on the board.

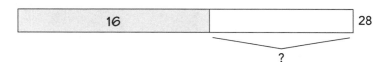

Explain: *With a continuous bar like this, we would draw a line to show the section of 16 eaten. The remaining piece, or difference, is the quantity that was not eaten. So we are trying to find the difference between 16 and 28. When we could see the squares we just counted them to find the difference. With a diagram like this we can think about using easy numbers, like 10s and 5s, to find the difference. Think of the bar almost like a number line. For example, if we go up 10 more from 16, we know that would be 26. I will write that on top of that section. That leaves a section with 2 more to get to 28. The 10 section + the 2 section makes 12, which is the difference.*

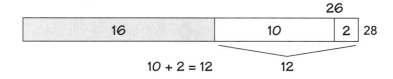

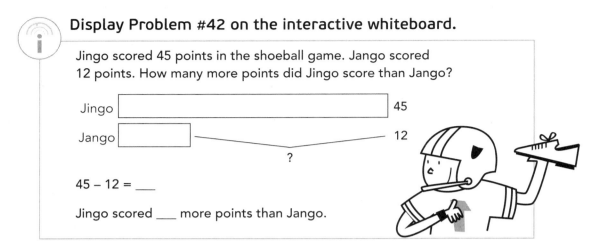

Display Problem #42 on the interactive whiteboard.

Jingo scored 45 points in the shoeball game. Jango scored 12 points. How many more points did Jingo score than Jango?

Jingo [ ] 45

Jango [ ] 12

?

45 – 12 = ___

Jingo scored ___ more points than Jango.

Read aloud the problem together, then ask children what kind of subtraction problem this is. (A comparison problem)

Say: *In this problem we are comparing two amounts and deciding how much bigger one is than the other. Remember that when we have a comparison problem, we want to put one amount above the other for easier comparing. As you can see on the board and your paper, we have two bars, one for each quantity. Let's label the bars with numbers, names, and an arrow and question mark to show the unknown.*

Call on a volunteer to model this on the board while other children do the same on their papers.

Explain: *We could begin by drawing a line to show where 12 is on the top bar. We know where it is on the bottom bar so we could just extend that line like this and label that top section 12. This is using what we know.* Demonstrate this on the board (see below).

Say: *Now we have to find the difference between 12 and 45. We could count up toward 45, like we did in the last problem, but let's try and use easy numbers to do that. If we go up 10 from 12, that would take us to 22. Up 10 more would get us to 32, and another 10 to 42. Now we're only 3 away from 45. Let's add up those sections: 10 + 10 + 10 + 3 = 33. The difference is 33.*

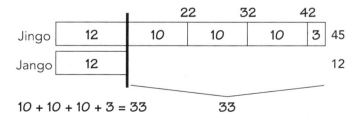

This efficient strategy is very similar to counting on a hundred chart or number line by tens and ones.

Have children work on Problems #43 and 44 (page 77) independently or in pairs. Remind them to draw bars to represent the amounts in each problem. Give children a few minutes to work. Then display each problem on the whiteboard. Call on volunteers to draw and label the diagram on the board. Then invite children to share their strategies for solving the problems, interacting on the whiteboard whenever possible.

Subtraction With Regrouping

Materials: student pages 78–79, pencils, projector, interactive whiteboard, markers

Preparation: Distribute copies of pages 78–79 and pencils to children. Go to www.scholastic.com/problemsolvedgr2 and click on Lesson 12. Set up your computer and projector to display the problems on the interactive whiteboard.

As they continue to subtract larger numbers, children become more adept at drawing and using continuous bars to represent the quantities. Remind children to use labels and an arrow and question mark for the unknown quantity. The problems in this lesson feature equations that require regrouping.

Checking With the Traditional Algorithm

As children gain more experience with Bar Modeling, they may find it helpful and reassuring to check their bar model work by using the traditional vertical algorithm. The reverse could also be done, where children check or prove their algorithm work by drawing an accompanying bar model. If the two match up, they most likely have solved the problem correctly.

Display Problem #45 on the interactive whiteboard.

Planet Zebo has 65 spaceships. Planet Zing has 27 spaceships. How many more spaceships does Planet Zebo have than Planet Zing?

Zebo [   ] 65

Zing [   ] 27

?

65 – 27 = ___

Planet Zebo has ___ more spaceships than Planet Zing.

Read aloud the problem and have children identify the facts and question.
Say: *This problem does not have a diagram so we will have to draw it. How many bars do we need to draw?* (2) *Why?* (We're comparing two things.)

Have children draw the bars on their papers, reminding them to draw the bars in proportion to each other and to label them with names and quantities. Call on a volunteer to model on the board, guiding the student, if necessary, to make the second bar about half the size of the first.

Say: *We want to find out how many more spaceships Zebo has than Zing. So let's draw a line to show where 27 is on the top bar. We could just extend the line from the bottom bar to the top bar and label that top section 27. Then, let's draw an arrow and question mark to show the unknown. Now we can find the difference between 65 and 27. Work with a partner to see if you can find the unknown quantity. Remember, you can use a counting-up strategy. Look for 10s and 5s that could help.*

Give children a few minutes to work, then call on volunteers to share their strategies and diagrams on the board and discuss.

Explain: *One possible strategy is to section a 3 next to the 27. This would bring us up to 30. From there we can count by 10s to 60. Then we just add 5 more to get to 65. This gives us 10 + 10 + 10 + 3 + 5 = 38.*

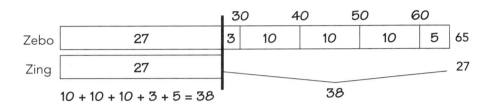

Another interesting way would be to count by 10s to 67—that would be 4 tens or 40. That is too high by 2, so we need to subtract 2 to get to 38. This is a more sophisticated mental strategy that could also be diagrammed with bars (see below), but some children may like it.

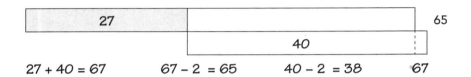

((i)) Display Problem #46 on the interactive whiteboard.

There were 81 birds sitting on giant robot. When he sneezed, 66 birds flew off. How many birds were still on the giant robot?

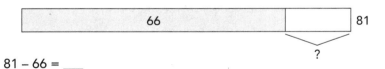

81 – 66 = ___

There were ___ birds still on the giant robot.

Read aloud the problem and have children identify and mark the facts and question on their papers.

Ask: *How many bars do we need to draw to represent this problem?* (1) *Why?* (It's a take-away problem.)

Say: *Let's draw a bar to represent the 81 birds and label its value.* Model drawing the bar on the board while children do the same on their papers.

Ask: *So what happened to some of the birds in the problem?* (They flew away.) *Let's draw a line to cut the bar and show a section of birds that flew away. How many flew away?* (66) *Since that's more than half, let's draw a line here. We'll shade that section in and label it with* 66.

Say: *Next, let's draw an arrow and question mark under the section that's not shaded. What does this represent?* (The birds that are still on the robot)

Have children work in pairs to find the unknown quantity. Remind them they can use a counting-up strategy and look for 10s and 5s that could help.

If needed, share the diagram below as one way to solve the problem. Explain that we can move up to the nearest 10 from 66 to 70 by adding 4. From 70 to 81 is 11 more. So 4 + 11 = 15, which is the difference.

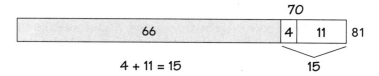

It is also possible to work backwards from 81 and count down. If we subtract 1 from 81 we are at 80. Take 10 more off and we are at 70. Count down 4 more and we get to 66. So 1 + 10 + 4 = 15, which is the difference.

Have children work on Problems #47 and 48 (page 79) independently or in pairs. Remind them to draw bars to represent the amounts in each problem. Give children a few minutes to work. Then display each problem on the whiteboard. Call on volunteers to draw and label the diagram on the board. Then invite children to share their strategies for solving the problems, interacting on the whiteboard whenever possible.

Bar Modeling With Addition and Subtraction (Within 1,000)

As children tackle problems with even larger numbers, Bar Modeling becomes even more helpful in solving the problems. Bar models can be used in conjunction with place value as an effective problem-solving strategy.

Continue to follow the same lesson routine as before: read aloud problems together, go through the problem-solving process (identify and underline the facts and circle the question), and share and discuss solutions. Have children work in pairs as much as possible.

LESSON 13

Addition With Larger Numbers

Materials: student pages 80–81, pencils, projector, interactive whiteboard, markers

Preparation: Distribute copies of pages 80–81 and pencils to children. Go to www.scholastic.com/problemsolvedgr2 and click on Lesson 13. Set up your computer and projector to display the problems on the interactive whiteboard.

The problems in this lesson do not include diagrams or equations. Children will have to draw and label their own bar diagrams, as well as write their own equations to match their work. Emphasize to children the importance of labeling their diagrams with numbers and names, as well as with the arrow and question mark for the unknown quantity.

Display Problem #49 on the interactive whiteboard.

Fizbop flew her rocket 155 miles to Planet Pony.
She then flew 165 miles to Planet Pineapple.
How many miles did she fly in all?

155
165
?

Fizbop flew ____ miles in all.

Read aloud the problem together, then ask children to identify the facts and question. Have children draw a bar model diagram on their papers to represent the amounts in the problem.

Say: *Since the amounts are fairly close the bars should be pretty close in length also. Make sure to label your diagram and include an arrow and a question mark to show the unknown. And remember to write an equation, or number sentence, to match your work.*

Have children work in pairs to solve the problem. Then call on volunteers to share their diagrams and solutions on the board and discuss.

One useful strategy to share with the class is to section these bars into hundreds, tens, and ones.

Explain: *In earlier problems, we tried to combine and make tens. As we get into larger numbers, we can do the same and look for hundreds to combine as well as find ways to make 100s. In this problem we can section the top bar into 100 + 50 + 5. We can also section the bottom bar into 100 + 50 + 10 + 5. Now adding will be easy. We combine the 100s to make 200. We combine the 50s to make another 100. Now we have 300. All we have to do is add on the extra 10 and two 5s for a total of 320.*

100	50	5	155

100	50	10	5	165

> 320

100 + 100 = 200   50 + 50 = 100   5 + 5 = 10

300 + 20 = 320

With these additions of 100s, 50s, and 10s, children could write out the equations each time or do them mentally. You may want to encourage the mental work, and then when debriefing write out these thoughts on the board as children share them orally.

Display Problem #50 on the interactive whiteboard.

Addie made 215 hot-pepper muffins. Her brother Arnold made 381. Her sister Argyle made 199. How many hot-pepper muffins did they make in all?

Ad. _____ 215

Arn. _____ 381 ?

Arg. _____ 199

They made ____ hot-pepper muffins in all.

Read aloud the problem and have children identify the facts and question.
Say: *In this problem we have three amounts. How many bars do we need to draw?* (3)

Have children draw bars on their papers to represent the three amounts and label as usual with an arrow and question mark to show the unknown. Two bars will be close in length and the other almost twice as long. Remind children to write an equation or equations to match their work. Have children work in pairs to solve, then invite them to share solutions on the board and discuss.

Explain: *Even with three addends, it makes sense to section the bars into hundreds, tens, and ones and to combine amounts to create these when possible. First let's add the hundreds: 200 + 300 + 100 = 600. Now let's see if we can make another hundred. We can by combining the 90 + 10. That makes 700 now. If we add 80 to that we have 780. You may notice we have a 1 + 9 to make another 10. That's 790. Add on the extra 5, and we have 795.*

Some children can do most of this mentally, but allow others to check off amounts as they add them up.

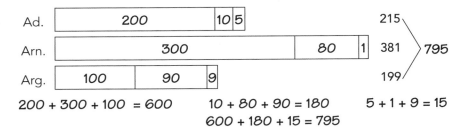

$$200 + 300 + 100 = 600 \qquad 10 + 80 + 90 = 180 \qquad 5 + 1 + 9 = 15$$
$$600 + 180 + 15 = 795$$

Have children work on Problems #51 and 52 (page 81) independently or in pairs. Remind them to draw bars and write equations to represent each problem. Give children a few minutes to work. Then display each problem on the whiteboard. Call on volunteers to draw and label the diagram on the board. Then invite children to share their strategies for solving the problems, interacting on the whiteboard whenever possible.

LESSON 14

Subtraction With Larger Numbers

Materials: student pages 82–83, pencils, projector, interactive whiteboard, markers

Preparation: Distribute copies of pages 82–83 and pencils to children. Go to www.scholastic.com/problemsolvedgr2 and click on Lesson 14. Set up your computer and projector to display the problems on the interactive whiteboard.

The focus in this lesson is on subtraction with larger numbers. Children will draw their own bar model diagrams and label them, as well as write equations to match their work.

Display Problem #53 on the interactive whiteboard.

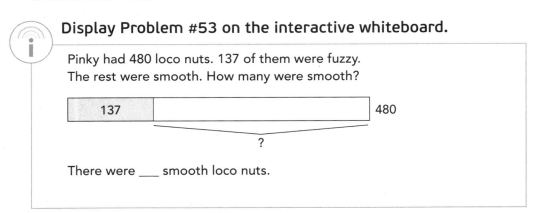

Pinky had 480 loco nuts. 137 of them were fuzzy. The rest were smooth. How many were smooth?

There were ___ smooth loco nuts.

Read aloud the problem together and ask children to identify and mark the facts and question on their papers.

Ask: *What kind of subtraction problem is this?* (Take-away problem) *So how many bars do we need to draw for this kind of problem?* (1)

Have children draw the bar on their papers then draw a line to section off 137 and shade it in. Remind children to label the bar with values and to draw an arrow and question mark to show the unknown. Afterwards, have children work in pairs to solve the problem. Then call on volunteers to share their solutions on the board.

Explain: *Once again, we can count up as if on a number line to find the difference. We want to look for hundreds, tens, and ones. In this case, adding 3 to 137 brings us to our first ten, 140. This is a good opportunity to use a new idea. Instead of adding 60 on to make 200, we could add 100 to get to 240. Keep going up by 100s to 340, then 440. Now we only have 40 left to get to 480. Let's add our jumps: 100 + 100 + 100 + 40 + 3 = 343.*

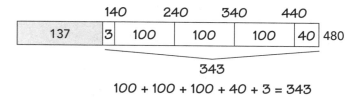

$$100 + 100 + 100 + 40 + 3 = 343$$

Display Problem #54 on the interactive whiteboard.

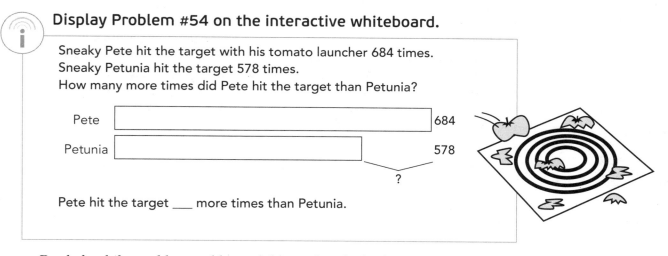

Sneaky Pete hit the target with his tomato launcher 684 times.
Sneaky Petunia hit the target 578 times.
How many more times did Pete hit the target than Petunia?

Pete hit the target ___ more times than Petunia.

Read aloud the problem and have children identify the facts and question.

Say: *Here we have a comparison problem. So we need two bars to show both quantities. On your paper draw two bars, one above the other, to show the quantities. Label the bars with numbers and names, then draw an arrow and question mark for the unknown. Then write a matching equation.*

Remind children that a good way to see the difference between the two bars is to draw a line from the end of the 578 bar up through the 684 bar. This shows us where the 578 is on the top bar and helps us see exactly how much of the top bar is longer than the bottom bar.

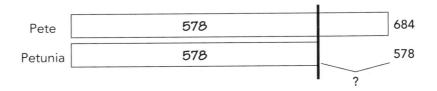

Silent Modeling

One fun and effective method of sharing a strategy like the one described in Problem #54 is to tell children: *I am going to model my thinking, but I am not going to say anything. See if you can figure out what I am doing.* Do each step slowly and very deliberately on the board. Pause after each move and stand aside for a moment. At the end of your demonstration, ask children to share what they think you were doing.

Explain: *Now we can begin to count up from 578 to 684 to find the difference. We could go up 100 right away to 678. That's pretty close. Add 2 more to get to the nearest ten; that's 680. Then we add 4 more to get to 684. Our addends look like this: 100 + 2 + 4 = 106.*

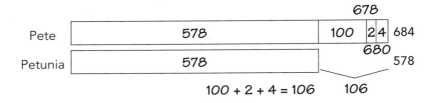

There are many other combinations of logical addends to find the solution. Invite children to share any other strategies on the board and discuss.

Have children work on Problems #55 and 56 (page 83) independently or in pairs. Remind them to draw bars and write equations to represent each problem. Give children a few minutes to work. Then display each problem on the whiteboard. Call on volunteers to draw and label the diagram on the board. Then invite children to share their strategies for solving the problems, interacting on the whiteboard whenever possible.

Bar Modeling With Basic Multiplication Problems

At this point children have become experts at using Bar Modeling in addition and subtraction problems. In this chapter we will investigate ways Bar Modeling can be useful in solving problems that involve multiplication.

Continue to follow the same lesson routine as before: read aloud problems together, go through the problem-solving process (identify and underline the facts and circle the question), and share and discuss solutions. As much as possible, have children work in pairs.

LESSON 15

Repeated Addition and Multiplication

Materials: student pages 84–85, pencils, projector, interactive whiteboard, markers

Preparation: Distribute copies of pages 84–85 and pencils to children. Go to www.scholastic.com/problemsolvedgr2 and click on Lesson 15. Set up your computer and projector to display the problems on the interactive whiteboard.

To ease into Bar Modeling with multiplication, children will begin with repeated addition problems. This will help them understand that repeated addition is the basis of multiplication. Instead of stacking multiple bars one on top of the other (as with addition), children will use a horizontal format to represent all the addends—a more efficient format for multiplication. The arrow and question mark indicating the unknown will now go under the diagram. So 3 × 5, for example, would look like the bar model below.

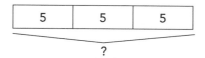

Display Problem #57 on the interactive whiteboard.

Susie picked 3 fuzzy flowers every day for 4 days.
How many fuzzy flowers did she pick in all?

| 3 | 3 | 3 | 3 |

?

3 + 3 + 3 + 3 = ___   OR   4 × 3 = ___

Susie picked ___ flowers in all.

Read aloud the problem together and have children identify and mark the facts and question on their papers.

Explain: *Here we have a problem where the same quantity is added again and again. Susie picked 3 flowers every day for 4 days. She got the same amount 4 times. Notice that instead of four bars, we have one long bar that's divided into 4 sections. Each group of 3 flowers is represented by a section of the bar. There are 4 sections, and each section is worth 3. The arrow and question mark show that we are trying to determine how much the sections are worth altogether.*

Have children work in pairs to solve the problem. Then invite volunteers to share their solutions on the board. Some children may know the 4 × 3 multiplication fact, while others may skip count or add by 3s or see that 3 + 3 = 6 and 6 + 6 = 12.

Explain: *The multiplication sentence says 4 × 3, which also means 4 groups of 3. Since there were 4 days and 3 flowers were picked on each day, we see that 4 × 3 matches that situation.*

 **Display Problem #58 on the interactive whiteboard.**

Cowboy Bob has 5 ponies. Each pony has 2 saddles.
How many saddles are there in all?

2	2	2	2	2

?

2 + 2 + 2 + 2 + 2 = ___ OR 5 × 2 = ___

There are ___ saddles in all.

Read aloud the problem and ask children to identify the facts and question.

Say: *Just like in the last problem, we have a situation where the same quantity is added again and again. Let's represent this with a bar model. What are we trying to find out?* (How many saddles are there in all?)

Ask: *How many ponies were there?* (5) *So let's draw a bar and section it into 5 parts—one part for each pony.* Draw this on the board and have children do the same on their papers.

Ask: *How many saddles does each pony have?* (2) *So let's label each section of the bar with a value of 2. Now we have 5 sections or groups of 2.* Continue modeling this on the board as children follow on their papers.

Say: *So we are trying to find out the total value of this bar with all the sections added together.* Draw the arrow and question mark under the bar.

Have children work in pairs to solve the problem and share solutions. Some children will know the 5 × 2 multiplication fact. Some may skip count or add by 2s.

Explain: *The multiplication sentence says 5 × 2, which also means 5 groups of 2. Since there were 5 ponies and each pony had 2 saddles, we see that 5 × 2 matches that situation.*

Have children work on Problems #59 and 60 (page 85) independently or in pairs. Remind them to draw bars to represent the amounts in each problem. Give children a few minutes to work. Then display each problem on the whiteboard. Call on volunteers to draw and label the diagram on the board. Then invite children to share their strategies for solving the problems, interacting on the whiteboard whenever possible.

LESSON 16
Basic Multiplication

Materials: student pages 86–87, pencils, projector, interactive whiteboard, markers

Preparation: Distribute copies of pages 86–87 and pencils to children. Go to www.scholastic.com/problemsolvedgr2 and click on Lesson 16. Set up your computer and projector to display the problems on the interactive whiteboard.

This lesson continues to help children think about repeated addition as the basis of multiplication. The diagram format will again be horizontal rather than one bar on top of the other. It is more efficient this way, just as multiplication is more efficient than repeated addition.

Display Problem #61 on the interactive whiteboard.

Chef Cherise baked 6 avocado brownies. She put 5 olives on top of each brownie. How many olives did she use in all?

?

5 + 5 + 5 + 5 + 5 + 5 = ___   OR   6 × 5 = ___

Chef Cherise used ___ olives in all.

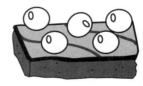

Read aloud the problem and have children identify the facts and question.
Say: *We have another situation where the same quantity is added again and again. Let's represent this with a bar model. What are we trying to find out?* (How many olives did Cherise use?)
Ask: *How many brownies are there?* (6) *Let's draw a bar and section it into 6 parts, one part for each brownie.* Draw this on the board and have children do the same on their papers.
Ask: *How many olives did each brownie have?* (5) *Let's label each section of the bar with a value of 5. Now we have 6 sections, or groups, of 5.* Do this on the board as children follow on their papers.
Say: *So we are trying to find out the total value of this bar with all the sections added together.* Draw the arrow and question mark under the bar to indicate the unknown.

Have children work in pairs to solve the problem and share their strategies and solutions. Some children may know the 6 × 5 multiplication fact. Some may count by 5s or make groups of 10. Have children fill in the multiplication sentence with their answer.

Explain: *The multiplication sentence says 6 × 5, which also means 6 groups of 5. Since there were 6 brownies and each one had 5 olives, we see that 6 × 5 matches that situation.*

Display Problem #62 on the interactive whiteboard.

There were 4 children in Jungle Jane's Gym Class. Each student did 8 push-ups. How many push-ups did they do altogether?

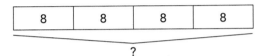

They did ___ push-ups altogether.

Read aloud the problem and have children identify the facts and question.

Say: *Here's another problem in which the same quantity is added again and again. We have 4 children, and each child did the same number of push-ups: 8. On your paper draw a bar model diagram to represent this problem. Label all the quantities and use an arrow and question mark to show the unknown. Then write an addition sentence and a multiplication sentence to match your diagram.*

When children have completed their diagrams, have them work in pairs to solve the problem. Then have them share their strategies and solutions with the class. Some children may know the 4 × 8 multiplication fact. Some may skip count by 8s, while others may add 8 + 8 = 16 and then add another 16 to that again.

Here is another strategy you might want to share with children: Split the sections into even smaller sections to create easier numbers, as we have done with addition and subtraction in the past. For example, if we decide to look for easier numbers such as 5s or 10s, we can split each section of 8 into 5 + 3 (see below). Doing this sectioning, we can then easily add the 5s to get 20, and then count up with the remaining 3s: 23, 26, 29, 32.

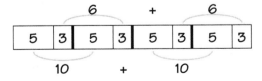

Have children work on Problems #63 and 64 (page 87) independently or in pairs. Remind them to draw bars and write equations to represent each problem. Give children a few minutes to work. Then display each problem on the whiteboard. Call on volunteers to draw and label the diagram on the board. Then invite children to share their strategies for solving the problems, interacting on the whiteboard whenever possible.

Bar Modeling With Measurement and Money Problems

In this chapter children will use Bar Modeling to solve problems about measurement and money. The problems involve addition and subtraction, so this should feel familiar to children. They will draw and label their own diagrams as well as write their own equations to match their diagrams.

Follow the same lesson routine established earlier: read aloud problems together, go through the problem-solving process (identify and underline the facts and circle the question), and share and discuss solutions. As much as possible, have children work in pairs.

LESSON 17

Measurement and Money Addition Problems

Materials: student pages 88–89, pencils, projector, interactive whiteboard, markers

Preparation: Distribute copies of pages 88–89 and pencils to children. Go to www.scholastic.com/problemsolvedgr2 and click on Lesson 17. Set up your computer and projector to display the problems on the interactive whiteboard.

Measurement and money are the contexts for the problems in this lesson, with a focus on addition.

Display Problem #65 on the interactive whiteboard.

Nelson Nock Nee had a bag of turtle food that weighed 187 pounds. He bought another bag of turtle food that weighed only 27 pounds. How many pounds of turtle food does Nelson have in all?

187

27   ?

Turtle Food

Nelson has ___ pounds of turtle food in all.

Read aloud the problem and ask children to identify the facts and question. **Say:** *In the last two lessons, we were adding the same quantity again and again. It was easier to keep all quantities in the same horizontal bar because they had the same values. But in this problem we're back to having two different quantities. So we will need to draw two bars to represent each quantity.*

Have children draw the diagrams on their papers, reminding them to label the bars and show the unknown quantity with an arrow and question mark. Then have them write an equation to match the problem as well. Call on volunteers to share their diagrams and equations on the board.

Explain: *So in combining or adding these two amounts, we might start by thinking about tens and ones, and even hundreds, since we have such large numbers. We could section the first bar into 100 + 80 + 7 and the second bar into 20 + 7. Now we can see that we have 80 and 20. Combining those would give us a second 100, so we now have 200. All that's left are two 7 sections. That adds to 14. So the answer is 214 pounds.*

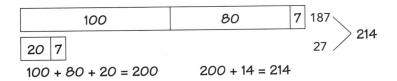

100 + 80 + 20 = 200   200 + 14 = 214

Tell children that when solving measurement problems, they should always include the unit of measurement in the answer.

Display Problem #66 on the interactive whiteboard.

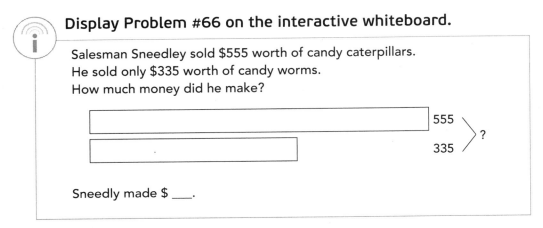

Salesman Sneedley sold $555 worth of candy caterpillars.
He sold only $335 worth of candy worms.
How much money did he make?

Sneedly made $ ___.

Read aloud the problem and have children identify the facts and question.
Say: *For this problem you will need to draw the two bars to represent each quantity. Label these bars and show the unknown with an arrow and question mark. Also, please write an equation to match the problem.*
   Call on a volunteer to draw and label the bar model on the board.
Say: *So in combining or adding these two amounts, we can start by thinking about hundreds, tens, and ones again. We could section the first bar into 500 + 50 + 5. The second bar could be sectioned into 300 + 30 + 5.*
   Allow children time to work on the problem in pairs. Then invite them to share their strategies and solutions on the board. Remind them to label the answer with the dollar sign.

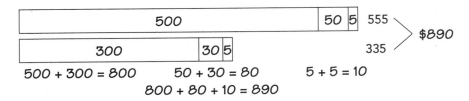

500 + 300 = 800   50 + 30 = 80   5 + 5 = 10
800 + 80 + 10 = 890

Have children work on Problems #67 and 68 (page 89) independently or in pairs. Remind them to draw bars and write equations to represent each problem. Give children a few minutes to work. Then display each problem on the whiteboard. Call on volunteers to draw and label the diagram on the board. Then invite children to share their strategies for solving the problems, interacting on the whiteboard whenever possible.

LESSON 18

Measurement and Money Subtraction Problems

Materials: student pages 90–91, pencils, projector, interactive whiteboard, markers

Preparation: Distribute copies of pages 90–91 and pencils to children. Go to www.scholastic.com/problemsolvedgr2 and click on Lesson 18. Set up your computer and projector to display the problems on the interactive whiteboard.

Measurement and money continue to be the contexts for the problems in this lesson, but with a focus on subtraction.

Display Problem #69 on the interactive whiteboard.

Cinderella sold a glass slipper for $211.
She spent $18 of that money on makeup.
How much money does she have left?

Cinderella has $____ left.

Read aloud the problem together and have children identify the facts and question. This is a good example of a problem that could be difficult for a second grader to do with the traditional algorithm as it involves two regroupings. With the bar model, we can solve without regrouping and also check that the algorithm works.

Ask: *What kind of subtraction problem do we have here—take away or comparison?* (Take away) *So how many bars do we need to draw to represent this problem?* (1)

Have children draw and label the bar diagram on their papers, reminding them to show the unknown with an arrow and question mark. Have them write a matching equation as well. Then call on volunteers to share their diagrams and equations on the board.

Document Camera Sharing

A document camera or other device that can project student work can be very helpful during these problem discussions. Having children draw their diagrams on their papers and then share via the projection system saves time and honors student work in a very visible way.

Explain: *In this problem the quantities are far apart, so we should consider big jumps. This case may lend us another look into an interesting strategy we explored earlier (Lesson 12). If we were to jump up 200 from 18, we would pass 211 and land at 218. We know we have gone too far, but we are only 7 too far. If we were to cut back 7 from our 200 jump that takes us right to 211. So 200 – 7 = 193, which is the difference between the numbers.*

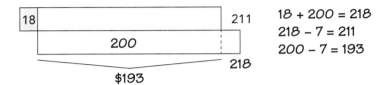

You can also discuss the traditional strategy of counting up, explaining that this strategy is like moving up a number line that ends in the total quantity. So in this case, children would count up 100 from 18 to get to 118. Add 2 more to get to 120 then go up 80 to get to 200. Add 11 more to get to 211. That gives us 100 + 2 + 80 + 11 = 193.

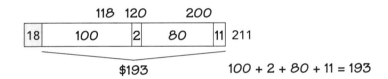

You may want to encourage children to notate numbers in the bar so they can keep track of the jumps up or the addends showing how far they've come with those jumps (see above diagram).

Display Problem #70 on the interactive whiteboard.

Tina the Giant needed 405 gallons of paint to paint her castle.
Ned the Giant needed 395 gallons of paint to paint his castle.
How much more paint did Tina need than Ned?

| T. | | 405 |
| N. | | 395 |

?

Tina needed ____ more gallons of paint than Ned.

Read aloud the problem and ask children to identify the facts and question.
Say: *In this problem we are looking at two different quantities being compared. How many bars do we need to draw to represent this problem?* (2)

Have children draw and label the bar diagram on their papers, reminding them to draw one bar above the other to represent the two different quantities. In this case, the two bars should be close in size. Have them also write a matching equation. Then call on volunteers to share their diagrams and equations on the board.

One strategy to share would be to draw a line showing where the 395 gallons in Ned's bar matches up with the 395 gallons in Tina's bar. Whatever is left over in Tina's bar is the difference. To determine exactly what that would be, children could count up. Just jumping up 5 would bring us to 400. Another 5 brings us to 405. So 5 + 5 = 10 gallons, which is the difference.

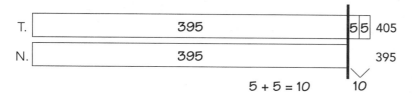

Have children work on Problems #71 and 72 (page 91) independently or in pairs. Remind them to draw bars and write equations to represent each problem. Give children a few minutes to work. Then display each problem on the whiteboard. Call on volunteers to draw and label the diagram on the board. Then invite children to share their strategies for solving the problems, interacting on the whiteboard whenever possible.

Bar Modeling With Multistep Problems

In this chapter children will combine everything they've learned about Bar Modeling so far to solve multistep problems involving addition, subtraction, and multiplication. Some problems may call for more than one bar model diagram, while others can be solved using only one diagram for all operations.

The lesson routine established earlier will be as valuable as ever: read aloud problems together, go through the problem-solving process to identify the facts and question, and share and discuss solutions. Have children work in pairs as much as possible, then invite volunteers to share their diagrams, equations, and strategies on the board.

LESSON 19

Mixed Multistep Problems

Materials: student pages 92–93, pencils, projector, interactive whiteboard, markers

Preparation: Distribute copies of pages 92–93 and pencils to children. Go to www.scholastic.com/problemsolvedgr2 and click on Lesson 19. Set up your computer and projector to display the problems on the interactive whiteboard.

In this lesson children will encounter multistep word problems that use addition and subtraction.

Display Problem #73 on the interactive whiteboard.

Lemona is making her famous lemonade. She mixes 55 gallons of water and 86 gallons of lemon juice. She puts 75 gallons of the mix in the refrigerator. She serves the rest.
How much lemonade does she serve?

Lemona serves ___ gallons of the lemonade.

Read aloud the problem and have children identify the facts and questions.
Say: *To solve this problem, we need to perform two operations. First we have to add, then we have to subtract. We can draw two bar model diagrams, one for each operation.*

Invite children to share ways to draw and section the bars for addition. For example, sectioning the bars into 50s and 5s can help: 50 + 50 = 100. 100 + 30 = 130. 130 + 11 = 141.

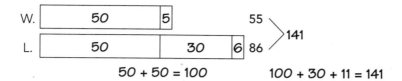

$$50 + 50 = 100 \qquad 100 + 30 + 11 = 141$$

Once we know that there are 141 gallons of lemonade we can draw a new bar with that value and subtract the 75 gallons that were put in the refrigerator. If we add up from 75 to 141 we can go up 25 to 100. 41 more brings us to 141: 25 + 41 = 66, which is the difference.

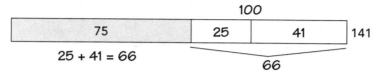

Say: *It is also possible to use just one diagram and complete all operations, addition and subtraction, on it.* Share the diagram below on the board.

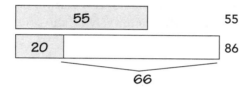

Explain: *In this diagram we draw both amounts in two bars and then shade out 55 from the top bar and 20 from the bottom bar. The 55 and 20 together make the subtracted 75. The unshaded part is 66, which is the answer.*

Most children will find Bar Modeling very handy in helping keep track of the various actions in the problem. The bar models serve as an easy reference and representation of these ideas. Some children may prefer using separate diagrams for each operation while others may like the combined approach.

Display Problem #74 on the interactive whiteboard.

> Gino made 61 peanut butter pizzas. He ate 42 of them.
> Then he bought 27 more later. How many pizzas does
> Gino have now?
>
>
> Gino has ___ pizzas now.

Read aloud the problem together and have children identify the facts and question. Point out that we can use two separate bar models to show the pizzas that were made and eaten. Encourage children to work in pairs to come up with strategies for drawing the bar models. Then invite them to share their approaches on the board.

For example, the first diagram could show the subtraction with an adding-up approach. Adding up 8 from 42 brings us to 50. 11 more brings us to 61. 8 + 11 = 19, which is the difference.

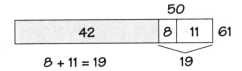

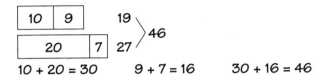

$8 + 11 = 19$

We then have to add the 19 and 27, which we can do with a two-bar diagram, as shown below.

$10 + 20 = 30$ $9 + 7 = 16$ $30 + 16 = 46$

Have children work on Problems #75 and 76 (page 93) independently or in pairs. Remind them to draw bars and write equations to represent each problem. Give children a few minutes to work. Then display each problem on the whiteboard. Call on volunteers to draw and label the diagram on the board. Then invite children to share their strategies for solving the problems, interacting on the whiteboard whenever possible.

LESSON 20

Challenging Multistep Problems

Materials: student pages 94–95, pencils, projector, interactive whiteboard, markers

Preparation: Distribute copies of pages 94–95 and pencils to children. Go to www.scholastic.com/problemsolvedgr2 and click on Lesson 20. Set up your computer and projector to display the problems on the interactive whiteboard.

Display Problem #77 on the interactive whiteboard.

Waldo spent $125 on a new tricycle and $116 on a helmet.
His Aunt Aloha gave him $179 to help pay for it.
How much money does Waldo still owe?

Waldo still owes $_____.

Read aloud the problem and have children identify the facts and question. **Say:** *Like in the last lesson, we can draw two bar-model diagrams to solve this multistep problem. In the first bar diagram, we can draw two bars, one above the other, to represent the two quantities. Then when we have the total, we can draw a second diagram for subtraction to find the difference between what Waldo got from Aunt Aloha and what he owes.*

Have children draw and label the diagrams on their papers. Then have them work in pairs to solve the problem. Afterwards, call on volunteers to share their diagrams and strategies on the board.

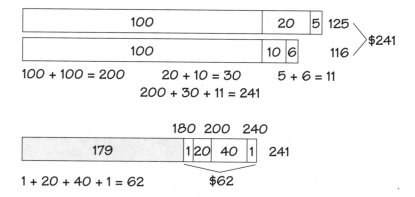

| 100 | 20 | 5 | 125 |

$241

| 100 | 10 | 6 | 116 |

100 + 100 = 200   20 + 10 = 30   5 + 6 = 11

200 + 30 + 11 = 241

180 200 240

| 179 | 1 | 20 | 40 | 1 | 241 |

1 + 20 + 40 + 1 = 62 $62

Display Problem #78 on the interactive whiteboard.

Leeza drove 21 miles each day on Monday, Tuesday, and Wednesday. On Thursday she drove 28 more miles. How many miles did she drive in all?

Leeza drove ___ miles in all.

Read aloud the problem and ask children to identify the facts and question. **Say:** *In this problem, Leeza drove the same amount over three days. What kind of operation do we need to do here?* (Multiplication or repeated addition)
Explain: *Remember, multiplication is best represented by one horizontal bar. We then split up the bar into three sections and label each section* 21. *After we've figured out how many miles Leeza drove over three days, we can draw another bar to add the 28 she drove on Thursday.*

Have children draw and label the bars on their papers, then work together in pairs to solve the problem.

3

| 20 | 1 | 20 | 1 | 20 | 1 |

60

20 × 3 = 60
1 × 3 = 3
60 + 3 = 63

| 60 | 3 | 63 |
| 20 | 8 | 28 |

91

60 + 20 = 80
8 + 3 = 11
80 + 11 = 91

Have children work on Problems #79 and 80 (page 95) independently or in pairs. Remind them to draw bars and write equations to represent each problem. Give children a few minutes to work. Then display each problem on the whiteboard. Call on volunteers to draw and label the diagram on the board. Then invite children to share their strategies for solving the problems, interacting on the whiteboard whenever possible.

Name _____

1. Bizzy found 9 bongo beetles.
Fizzy found 5 bongo beetles.
How many beetles did they find in all?

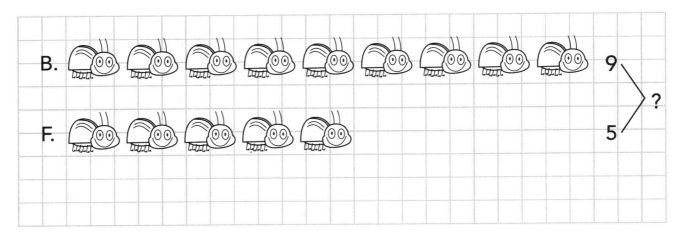

9 + 5 = _____

They found _____ bongo beetles in all.

2. Felix saw 4 robo-bunnies hopping down the street.
Then he saw 10 more robo-bunnies join them.
How many robo-bunnies are hopping now?

4 + 10 = _____

There are _____ robo-bunnies hopping now.

Name _____

3. Tina ate 10 purple peas. Then she ate 6 more purple peas.
How many purple peas did she eat altogether?

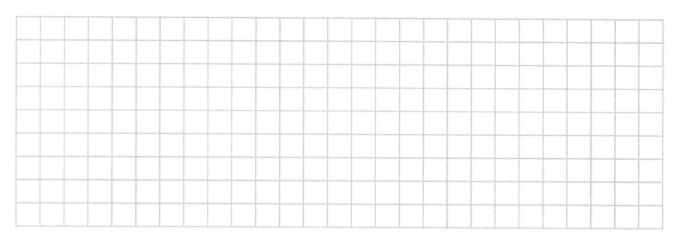

10 + 6 = _____

Tina ate _____ purple peas altogether.

4. Riz painted 5 spaceships green.
He painted 7 spaceships red.
How many spaceships did Riz paint in all?

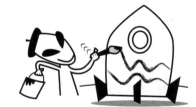

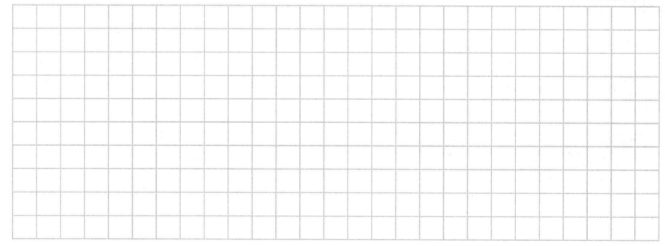

5 + 7 = _____

Riz painted _____ spaceships in all.

Name _____

5. Boris made 12 brownie bricks on Monday.
He made 4 brownie bricks on Tuesday.
How many brownie bricks did he make altogether?

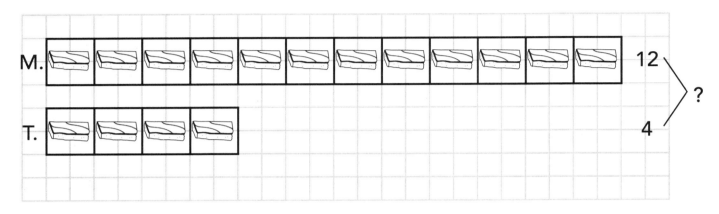

12 + 4 = _____

He made _____ brownie bricks altogether.

6. There were 3 astronauts at the space station.
The next day 9 more astronauts joined them.
How many astronauts are at the space
station now?

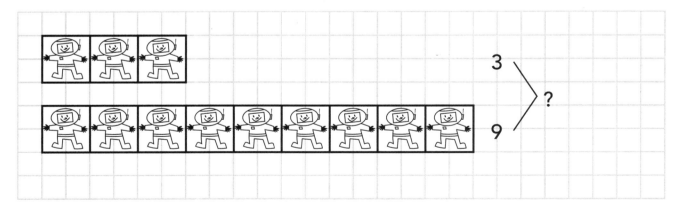

3 + 9 = _____

There are _____ astronauts at the space station now.

Name _____

7. Arnie the aardvark ate 11 chocolate ants. Amy the aardvark ate 6 chocolate ants. How many chocolate ants did they eat altogether?

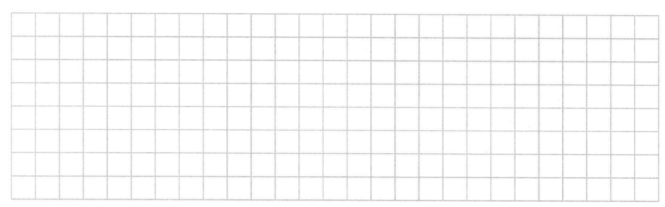

11 + 6 = _____

They ate _____ chocolate ants altogether.

8. There are 5 birds in the bed. There are 5 dogs in the sink. There are 5 cats in the tub. How many animals are there in all?

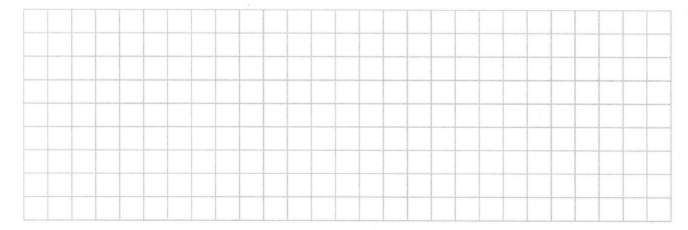

5 + 5 + 5 = _____

There are _____ animals in all.

Name _____

9. Ned placed 10 peanuts on his head. Ted placed 10 peanuts on his head. How many peanuts did the boys have on their heads altogether?

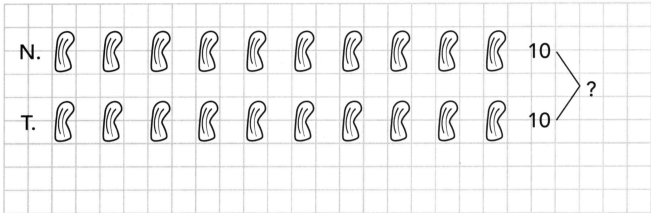

10 + 10 = _____

The boys had _____ peanuts on their heads altogether.

10. Billy Worm counted 13 birds. Willy Worm counted 5 birds. How many birds did they count altogether?

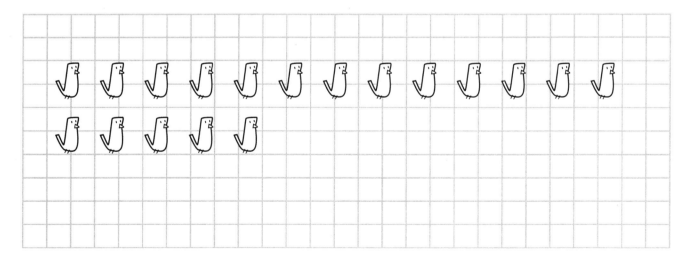

13 + 5 = _____

They counted _____ birds altogether.

Name _____

11. The teacher found 2 marmots in Stan's desk. She found 14 marmots in Jan's desk. How many marmots did the teacher find?

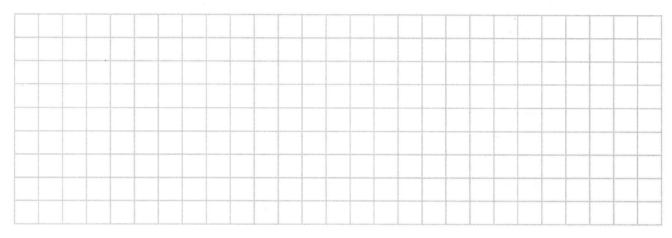

2 + 14 = _____

The teacher found _____ marmots.

12. Fargo wanted to get hats for all her friends. She had 7 friends that were girls and 8 friends that were boys. How many hats does Fargo need?

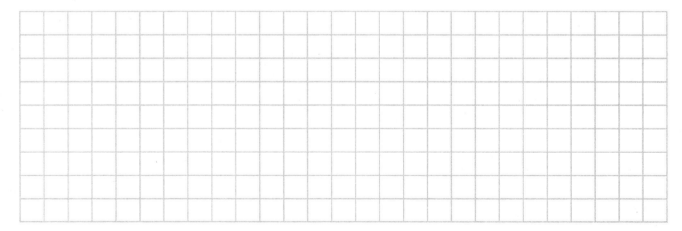

7 + 8 = _____

Fargo needs _____ hats.

13. Mandy sold 5 boxes of candy to the groundhogs.
She sold 11 boxes of candy to the squirrels.
How many boxes of candy did Mandy sell in all?

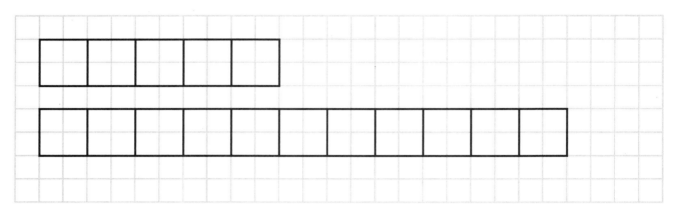

5 + 11 = _____

Mandy sold _____ boxes of candy in all.

14. 6 bats flew in the kitchen at night.
In the morning, 7 more bats joined them.
Then during the afternoon, 3 more bats flew in.
How many bats are in the kitchen altogether?

6 + 7 + 3 = _____

There are _____ bats in the kitchen altogether.

Name _____

15. Sammy ordered a dozen spinach donuts.
She also ordered 8 broccoli donuts.
How many donuts did Sammy order in all?

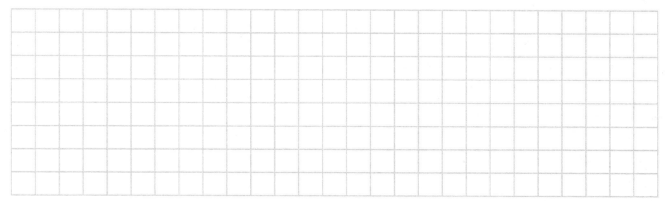

12 + 8 = _____

Sammy ordered _____ donuts in all.

---

16. Betty Bear ate 5 bowls of berry cereal for breakfast.
She ate 8 bowls of berry cereal for lunch.
How many bowls of berry cereal did Betty eat?

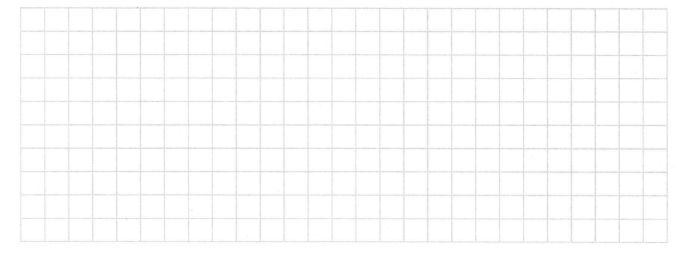

5 + 8 = _____

Betty ate _____ bowls of berry cereal.

17. Pez had 9 noodles. The Goo-Goo Monster
ate 5 of them. How many noodles does
Pez have left?

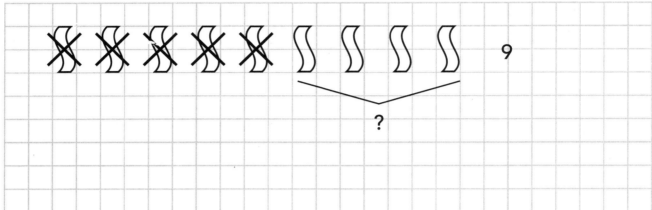

9 – 5 = _____

Pez has _____ noodles left.

18. There were 12 flying turtles. Then 9 flew away.
How many flying turtles were left?

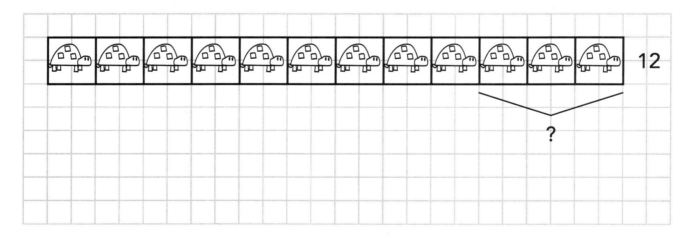

12 – 9 = _____

There were _____ flying turtles left.

19. Peanut had 12 silly stickers. He gave Lima Bean 3 stickers.
How many silly stickers are left?

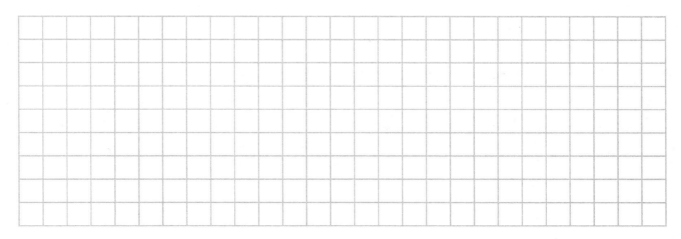

12 – 3 = _____

There are _____ silly stickers left.

20. Rhonda had 18 turkeys on her farm.
Then 16 went away to college.
How many turkeys are left on the farm?

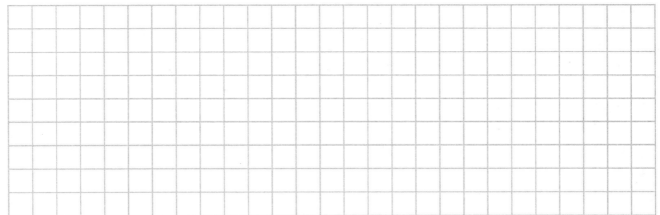

18 – 16 = _____

There are _____ turkeys left on the farm.

21. Clancy had 14 blue shoes. He had 8 purple shoes.
How many more blue shoes are there than purple shoes?

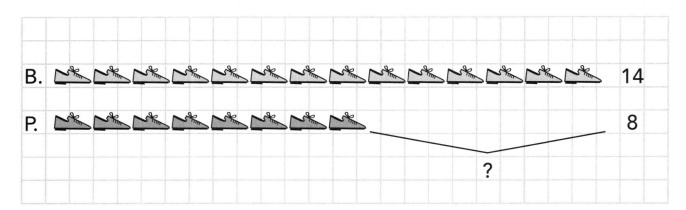

14 – 8 = _____

There are _____ more blue shoes than purple shoes.

22. Chef Crumb baked 2 spinach cakes.
She baked 17 cucumber cakes.
What is the difference between the number
of cucumber cakes and spinach cakes?

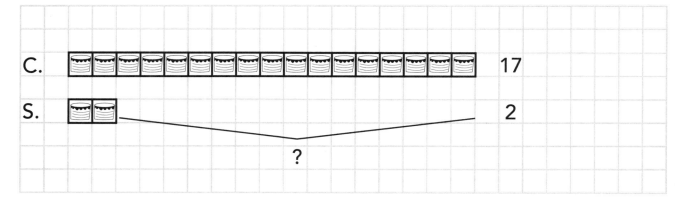

17 – 2 = _____

The difference between the number of spinach cakes
and cucumber cakes is _____.

Name _____

23. Carl the camel has 16 canteens of water.
He has 9 canteens of lemonade.
How many more canteens of water
are there than canteens of lemonade?

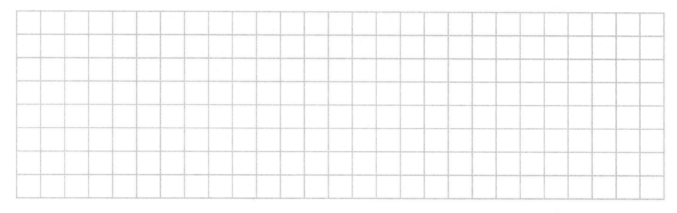

16 – 9 = _____

There are _____ more canteens of water than
there are of lemonade.

· ·

24. Cinderella had 12 magic mice who wore hats.
8 non-magic mice wore coats.
How many more magic mice were there than non-magic mice?

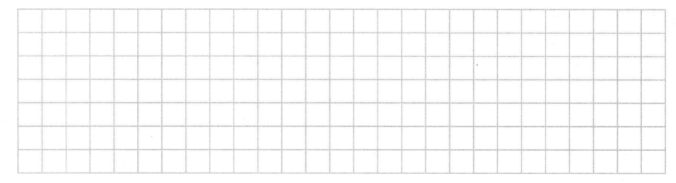

12 – 8 = _____

There were _____ more magic mice than non-magic mice.

25. Henrietta Hen will sing 20 songs at her concert. She has sung 10 songs so far. How many more songs does she have to sing?

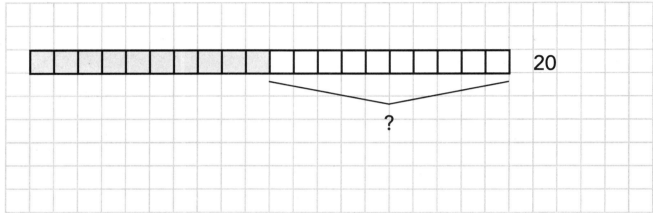

20 – 10 = _____

Henrietta has _____ more songs to sing.

26. Sammy Snake had 15 bowties. Sidney Snake had 7 bowties. What is the difference between their number of bowties?

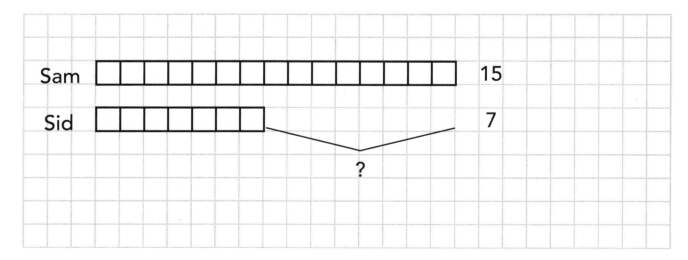

15 – 7 = _____

The difference between the number of bowties is _____.

Name _____

27. A cow had 14 Pookie Man cards. He lost 10 of them. How many Pookie Man cards does the cow have left?

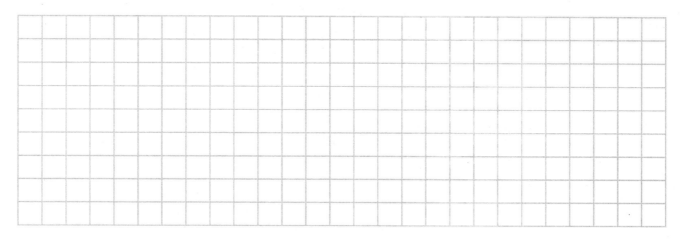

14 – 10 = _____

The cow has _____ Pookie Man cards left.

28. 20 dodgeball players went to a restaurant. 6 ordered hot dogs. The rest ordered hamburgers. How many dodgeball players ordered hamburgers?

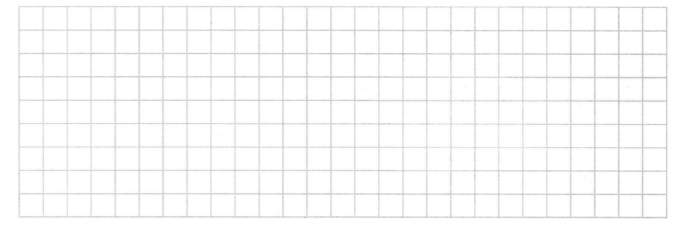

20 – 6 = _____

_____ dodgeball players ordered hamburgers.

29. Fantastic Freda had 16 pumpkin pops.
She gave 7 to her sister Fluffy. How many
pumpkin pops does Freda have left?

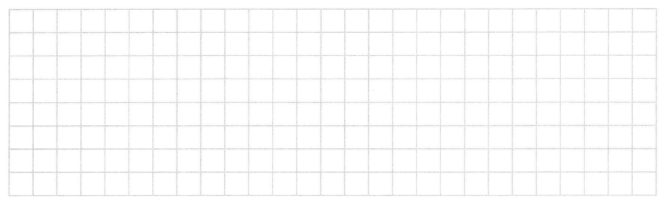

16 – 7 = _____

Freda has _____ pumpkin pops left.

30. Snappy scored 13 points in the gravity ball game.
Happy scored 8 points. How many more points
did Snappy score than Happy?

13 – 8 = _____

Snappy scored _____ more points than Happy.

Name _____

31. Sammy Stoneage had 20 rocks. He painted 11 yellow.
The rest he painted red. How many red rocks were there?

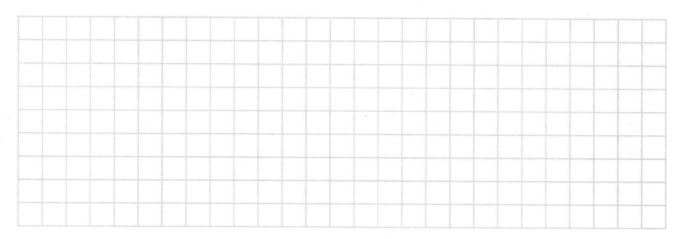

20 – 11 = _____

There were _____ red rocks.

32. Pina had 19 pickles. 12 had sugar on them.
How many of Pina's pickles did not have
sugar on them?

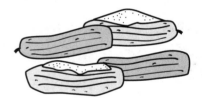

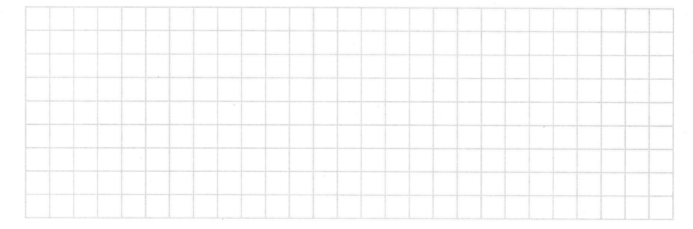

19 – 12 = _____

_____ of Pina's pickles did not have sugar.

Name _____

33. Mr. Cat R. Pillar has 22 socks.
He found 20 more under the bed.
How many socks does Mr. Pillar have now?

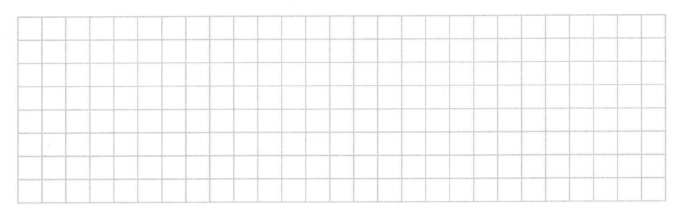

22 + 20 = _____

Mr. Pillar has _____ socks now.

34. Spot chased 40 elephants. He chased
24 giraffes. He chased 32 rhinos.
How many animals did Spot chase in all?

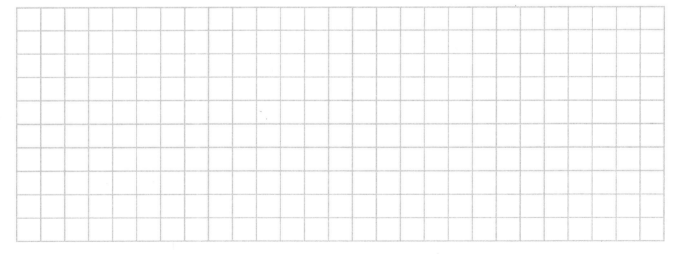

40 + 24 + 32 = _____

Spot chased _____ animals in all.

Name _____

35. Philly picked 45 golden mushrooms.
Milly picked 30 golden mushrooms.
How many golden mushrooms did they
pick altogether?

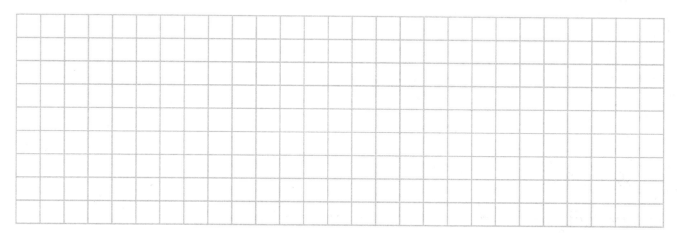

45 + 30 = _____

They picked _____ golden mushrooms altogether.

---

36. The turkey's hat is 30 years old.
How old will the hat be 30 years from now?

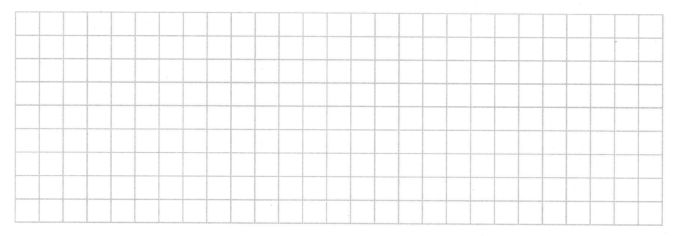

30 + 30 = _____

The hat will be _____ years old.

37. Ludmilla jumped 39 times over her brother Boris.
She jumped 61 times over her sister Olga.
How many times did Ludmilla jump altogether?

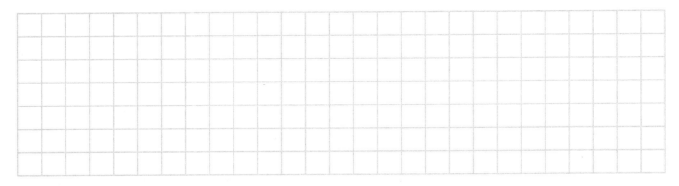

39 + 61 = _____

Ludmilla jumped _____ times altogether.

38. In the first round of the Baby Bottle Bowling
Championship, Rocky knocked down 88 bottles.
In the second round, he knocked down 15 more.
How many bottles did Rocky knock down in all?

88 + 15 = _____

Rocky knocked down _____ bottles in all.

Name _____

39. Manfred has 57 trained termites in his circus.
In June, 26 more termites joined the circus.
How many termites are in the circus now?

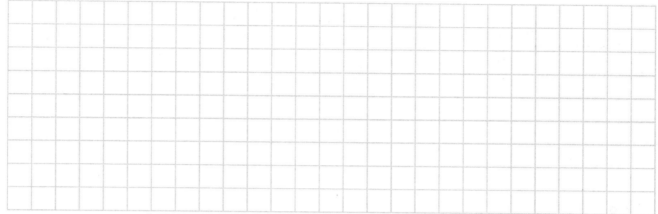

57 + 26 = _____

There are _____ termites in the circus now.

40. Zinnia grew 36 Snap Dragons in her garden. She also grew
55 Venus Fly Traps. How many plants did Zinnia grow?

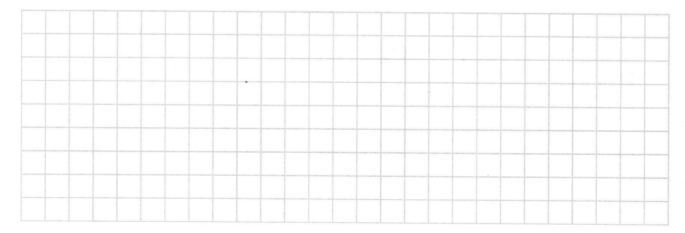

36 + 55 = _____

Zinnia grew _____ plants.

Name _____

41. There were 28 garlic gumdrops in the bag.
Max ate 16 of them. How many gumdrops were left?

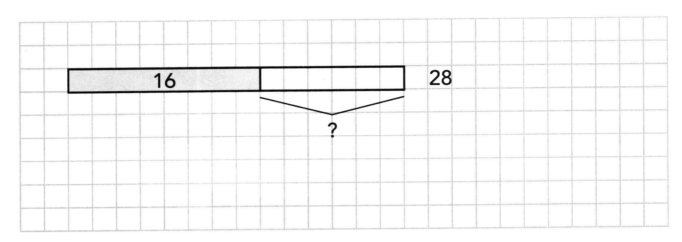

28 – 16 = _____

There were _____ garlic gumdrops left.

42. Jingo scored 45 points in the shoeball game.
Jango scored 12 points. How many more
points did Jingo score than Jango?

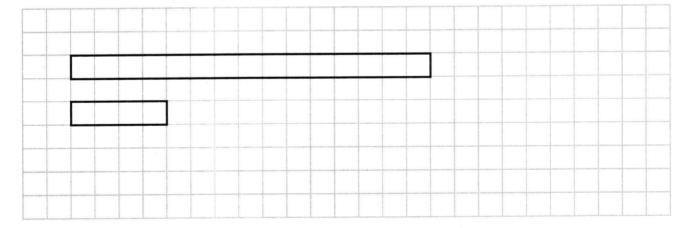

45 – 12 = _____

Jingo scored _____ more points than Jango.

Name _____

43. There were 36 pumpkins in the patch.
21 were purple. The rest were red.
How many pumpkins were red?

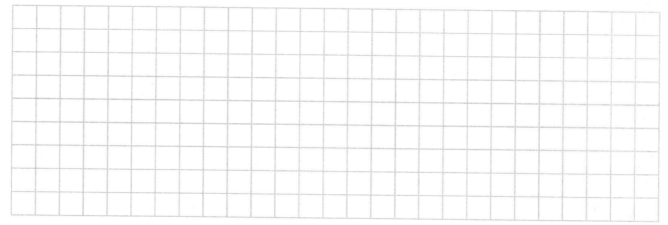

36 – 21 = _____

_____ pumpkins were red.

44. Slick had 59 tons of candy. He gave away 25 tons.
How many tons of candy does he have left?

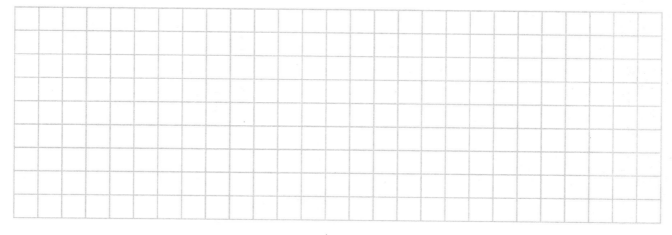

59 – 25 = _____

Slick has _____ tons of candy left.

Name _____

45. Planet Zebo has 65 spaceships. Planet Zing
has 27 spaceships. How many more spaceships
does Planet Zebo have than Planet Zing?

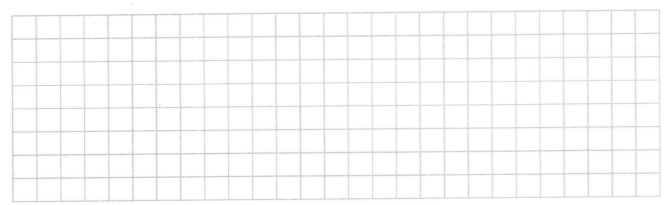

65 – 27 = _____

Planet Zebo has _____ more spaceships than Planet Zing.

46. There were 81 birds sitting on a giant robot.
When he sneezed, 66 birds flew off.
How many birds were still on the giant robot?

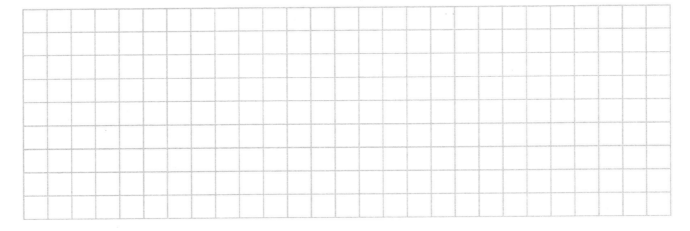

81 – 66 = _____

There were _____ birds still on the giant robot.

Name _____

47. There were 76 barrels of Jungle Juice. 18 barrels were empty. The rest were full. How many barrels were full?

76 – 18 = _____

There were _____ full barrels.

48. 50 monkeys went to the baseball game. There were only 38 seats left. How many monkeys did not get a seat?

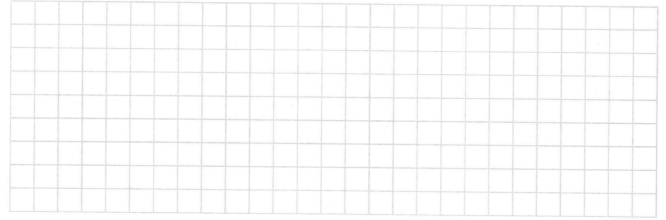

50 – 38 = _____

_____ monkeys did not get seats.

49. Fizbop flew her rocket 155 miles to Planet Pony.
She then flew 165 miles to Planet Pineapple.
How many miles did she fly in all?

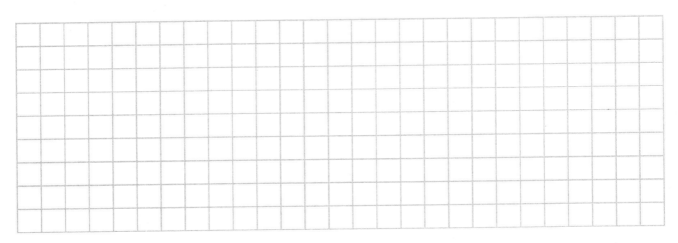

Fizbop flew _____ miles in all.

50. Addie made 215 hot-pepper muffins.
Her brother Arnold made 381.
Her sister Argyle made 199.
How many hot-pepper muffins did they make in all?

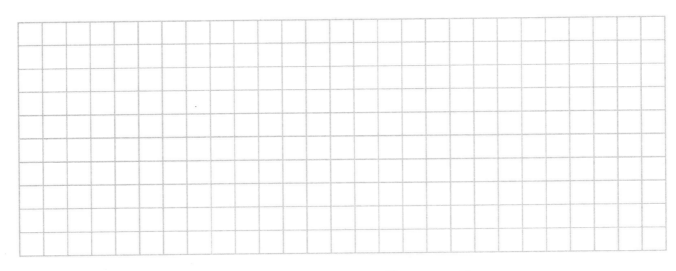

They made _____ hot-pepper muffins in all.

51. Happy the Clown bought 106 red noses for the clown carnival. Slappy the Clown bought 92 more. How many red noses did they buy in all?

They bought _____ red noses in all.

52. Slippery Sid collected 364 noodles last year.
He collected 518 noodles this year.
How many noodles does Sid have altogether?

Sid has _____ noodles altogether.

Name _____

53. Pinky had 480 loco nuts. 137 of them were fuzzy. The rest were smooth. How many were smooth?

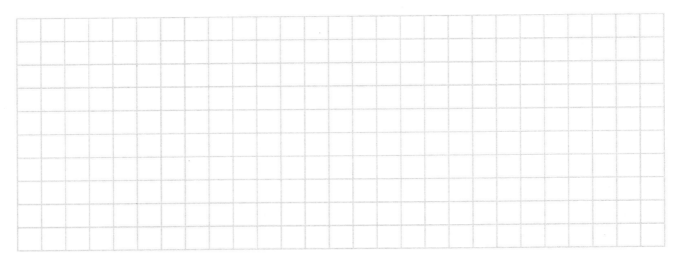

There were _____ smooth loco nuts.

54. Sneaky Pete hit the target with his tomato launcher 684 times. Sneaky Petunia hit the target 578 times. How many more times did Pete hit the target than Petunia?

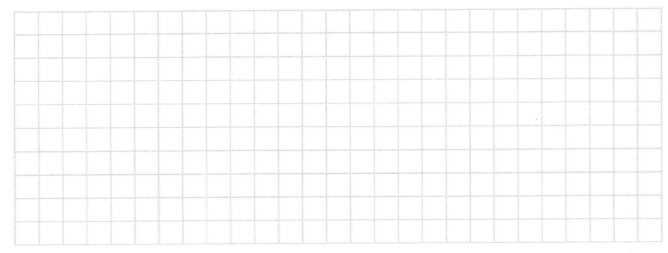

Pete hit the target _____ more times than Petunia.

Name _____

55. There were 505 chickens at Pat's
birthday party. 212 went home early.
How many chickens are still at the party?

There are _____ chickens still at the party.

56. Ringo had 135 feet of cloth to make a scarf for his giraffe.
He cut off 35 feet. How many feet of cloth were left?

There were _____ feet of cloth left.

57. Susie picked 3 fuzzy flowers every day for 4 days.
How many fuzzy flowers did she pick in all?

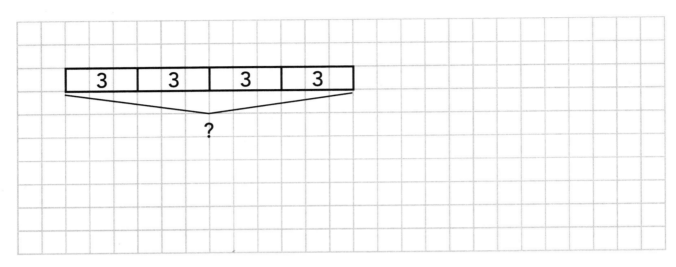

3 + 3 + 3 + 3 = _____   OR 4 × 3 = _____

Susie picked _____ flowers in all.

58. Cowboy Bob has 5 ponies.
Each pony has 2 saddles.
How many saddles are there in all?

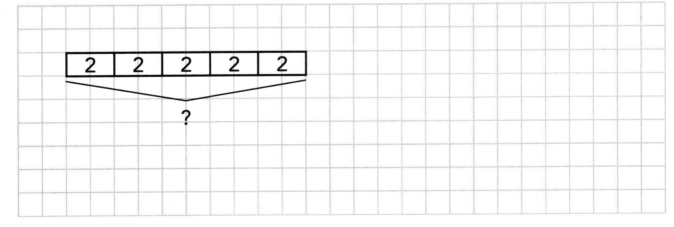

2 + 2 + 2 + 2 + 2 = _____   OR 5 × 2 = _____

There are _____ saddles in all.

Name _____

59. 4 Gazookians landed on Earth.
Each Gazookian had 4 eyes.
How many eyes did the Gazookians have in all?

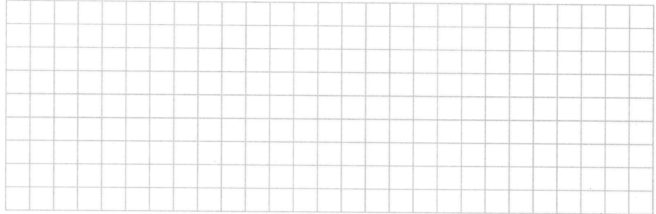

$4 + 4 + 4 + 4 =$ _____   OR $4 \times 4 =$ _____

The Gazookians had _____ eyes in all.

---

60. Pinky the Pirate found 2 treasure chests.
Each chest had 7 gold coins in it.
How many gold coins did Pinky find in all?

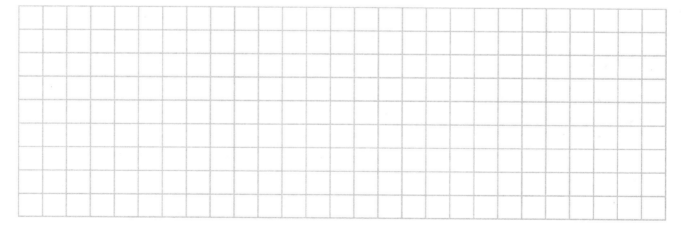

$7 + 7 =$ _____   OR $2 \times 7 =$ _____

Pinky found _____ gold coins in all.

61. Chef Cherise baked 6 avocado brownies.
She put 5 olives on top of each brownie.
How many olives did she use in all?

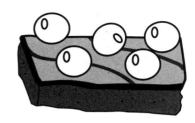

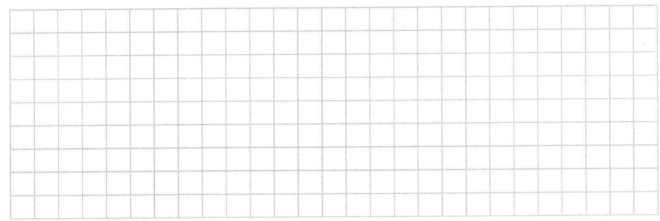

5 + 5 + 5 + 5 + 5 + 5 = _____ OR 6 × 5 = _____

Chef Cherise used _____ olives in all.

62. There were 4 students in Jungle Jane's Gym Class.
Each student did 8 push-ups.
How many push-ups did they do altogether?

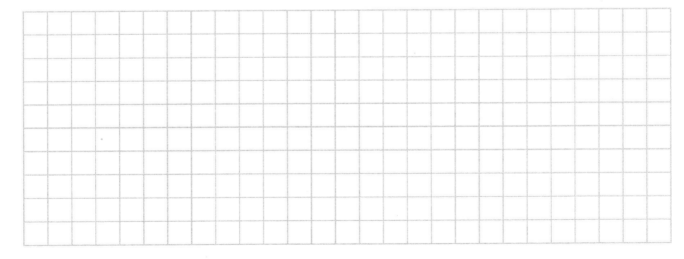

They did _____ push-ups altogether.

Name _____

63. Lucy played 3 games of Bingo Ball.
She scored 6 points in each game.
How many points did she score in all?

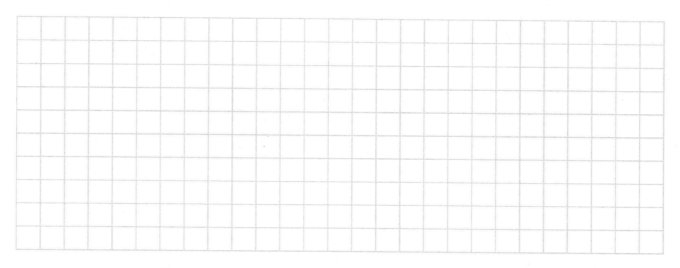

Lucy scored _____ points in all.

64. The Baloney Twins, Beth and Bill, each
had 9 salamanders under their bed.
How many salamanders did they have in all?

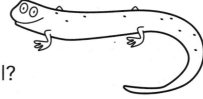

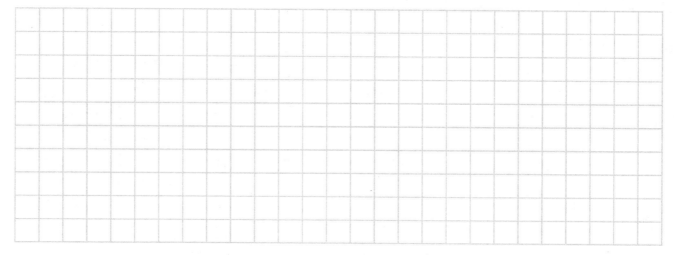

Beth and Bill had _____ salamanders in all.

Name _____

65. Nelson Nock Nee had a bag of turtle food
that weighed 187 pounds. He bought
another bag of turtle food that weighed only
27 pounds. How many pounds of turtle food
does Nelson have in all?

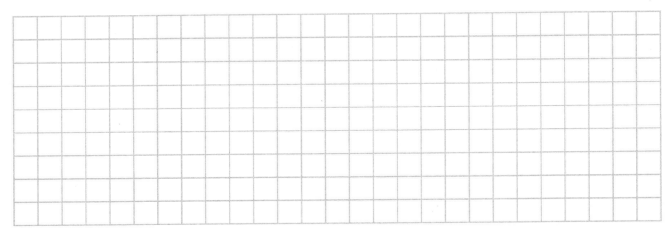

Nelson has _____ pounds of turtle food in all.

66. Salesman Sneedley sold $555 worth of candy caterpillars.
He sold only $335 worth of candy worms.
How much money did Sneedly make?

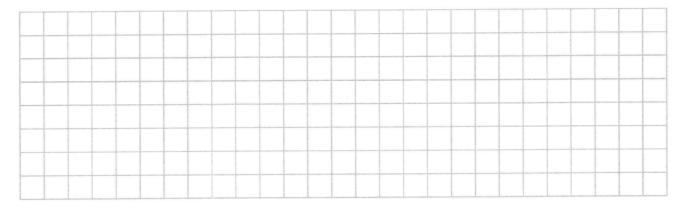

Sneedly made $ _____.

Name _____

67. Mrs. Funzers gave her class 215 minutes of recess in the morning.
She gave the class 119 minutes of recess in the afternoon.
How many minutes of recess did the class have altogether?

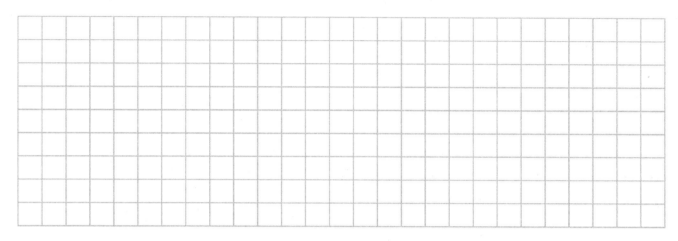

The class had _____ minutes of recess altogether.

68. Tinkerbell carried the football for 3 plays
during the game. She gained 18 yards,
21 yards, and 99 yards. How many yards
did she gain in all?

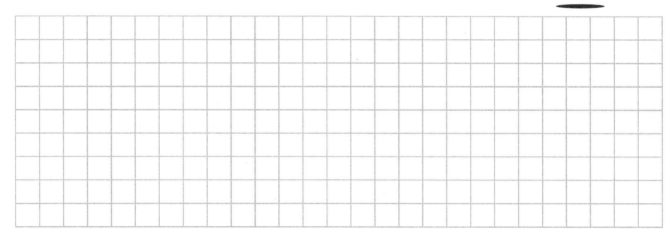

Tinkerbell gained _____ yards in all.

69. Cinderella sold a glass slipper for $211.
She spent $18 of that money on makeup.
How much money does she have left?

Cinderella has $_____ left.

70. Tina the Giant needed 405 gallons of paint to paint her castle.
Ned the Giant needed 395 gallons of paint to paint his castle.
How much more paint did Tina need than Ned?

Tina needed _____ more gallons of paint than Ned.

71. Ernie Explorer hiked 712 miles in 2 years in the Frisbee Forest. In the first year, he hiked 365 miles. How many miles did he hike in his second year?

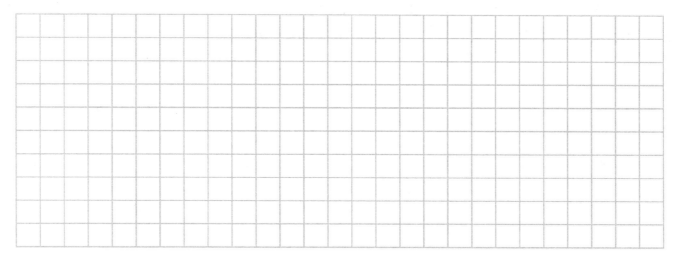

Ernie hiked _____ miles in his second year.

72. Ronald Rat had a piece of string cheese that was 150 inches long. He ate 76 inches of it. How long is the remaining string of cheese?

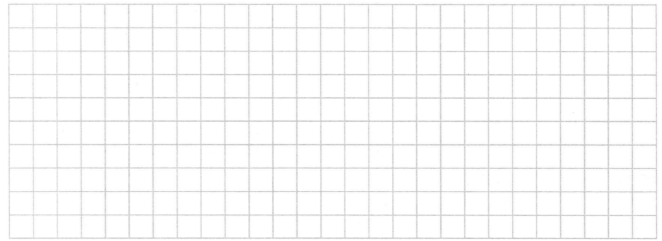

The remaining string of cheese is _____ inches long.

Name _____

73. Lemona is making her famous lemonade.
She mixes 55 gallons of water and 86 gallons
of lemon juice. She puts 75 gallons of the mix
in the refrigerator. She serves the rest.
How much of the lemonade does she serve?

Lemona serves _____ gallons of the lemonade.

---

74. Gino made 61 peanut butter pizzas. He ate 42 of them.
Then he bought 27 more later. How many pizzas
does Gino have now?

Gino has _____ pizzas now.

Name _____

75. There were 93 worms at a party. 83 more joined them. Then 58 worms left. How many worms were still at the party?

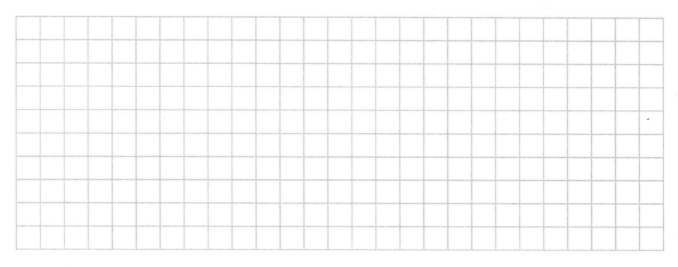

There were _____ worms still at the party.

76. Wilfred played 2 games of stink ball. He scored 45 points in each game. The referee gave him a penalty and he lost 17 points. How many total points did he finish with?

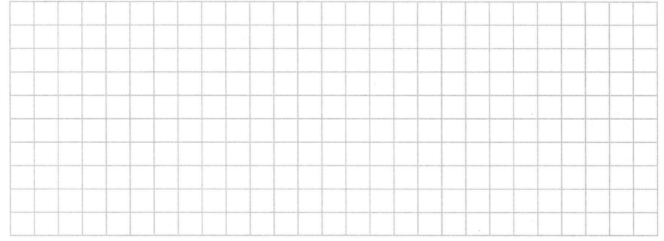

Wilfred finished with _____ total points.

Name _____

77. Waldo spent $125 on a new tricycle and $116 on a helmet.
His Aunt Aloha gave him $179 to help pay for it.
How much money does Waldo still owe?

Waldo still owes $_____.

78. Leeza drove 21 miles each day on
Monday, Tuesday, and Wednesday.
On Thursday, she drove 28 more miles.
How many miles did she drive in all?

She drove _____ miles in all.

Name _____

79. There were 465 people at the Happy Hippos
rock concert. After two songs, 329 left.
125 more people arrived after intermission
and stayed to the end. How many people
were there at the end of the concert?

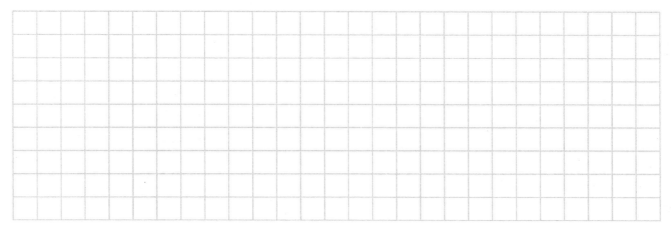

There were _____ people at the end of the concert.

---

80. 554 boys solved Problem #80. 250 girls solved Problem #80.
176 teachers solved Problem #80. How many more students
solved Problem #80 than teachers?

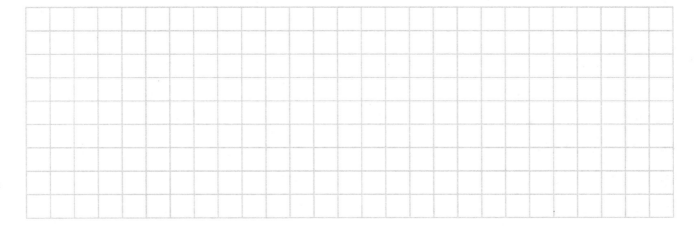

_____ more students solved Problem #80 than teachers.

Answer Key

1. 14 bongo beetles
2. 14 robo-bunnies
3. 16 purple peas
4. 12 spaceships
5. 16 brownie bricks
6. 12 astronauts
7. 17 chocolate ants
8. 15 animals
9. 20 peanuts
10. 18 birds
11. 16 marmots
12. 15 hats
13. 16 boxes of candy
14. 16 bats
15. 20 donuts
16. 13 bowls
17. 4 noodles
18. 3 flying turtles
19. 9 silly stickers
20. 2 turkeys
21. 6 more blue shoes
22. 15
23. 7 more water canteens
24. 4 more magic mice
25. 10 more songs
26. 8
27. 4 Pookie Man cards

28. 14 dodgeball players
29. 9 pumpkin pops
30. 5 more points
31. 9 red rocks
32. 7 pickles
33. 42 socks
34. 96 animals
35. 75 golden mushrooms
36. 60 years old
37. 100 times
38. 103 bottles
39. 83 termites
40. 91 plants
41. 12 garlic gumdrops
42. 33 more points
43. 15 red pumpkins
44. 34 tons
45. 38 more spaceships
46. 15 birds
47. 58 barrels
48. 12 monkeys
49. 320 miles
50. 795 hot-pepper muffins
51. 198 red noses
52. 882 noodles
53. 343 smooth loco nuts
54. 106 more times

55. 293 chickens
56. 100 feet of cloth
57. 12 flowers
58. 10 saddles
59. 16 eyes
60. 14 gold coins
61. 30 olives
62. 32 push-ups
63. 18 points
64. 18 salamanders
65. 214 pounds
66. $890
67. 334 minutes
68. 138 yards
69. $193
70. 10 more gallons
71. 347 miles
72. 74 inches
73. 66 gallons
74. 46 pizzas
75. 118 worms
76. 73 points
77. $62
78. 91 miles
79. 261 people
80. 628 more students